Fixed-Time Cooperative Control of Multi-Agent Systems

Zongyu Zuo · Qing-Long Han · Boda Ning

Fixed-Time Cooperative Control of Multi-Agent Systems

Springer

Zongyu Zuo
The Seventh Research Division
Beihang University (BUAA)
Beijing, China

Qing-Long Han
Swinburne University of Technology
Melbourne, VIC, Australia

Boda Ning
Swinburne University of Technology
Melbourne, VIC, Australia

ISBN 978-3-030-20278-1 ISBN 978-3-030-20279-8 (eBook)
https://doi.org/10.1007/978-3-030-20279-8

This Springer imprint is published by the registered company Springer Nature Switzerland AG
The registered company address is: Gewerbestrasse 11, 6330 Cham, Switzerland

Preface

A multi-agent system is a system that consists of a large group of geographically distributed and networked autonomous agents, where each of them represents a strategic entity capable of perceiving and computing scientific data from a physical world through sensors and acting upon a physical plant through actuators. Multi-agent systems, which have several advantages including autonomy, flexibility, manipulability, and scalability, can achieve coordinated objectives, such as consensus estimation, target enclosing, and formation tracking by communicating with each other through communication networks. There are a broad range of applications of multi-agent systems in mobile robots, exploration of hazardous environments, military surveillance and reconnaissance, search and rescue in disaster sites, spacecraft formation flying, and modern critical infrastructure monitoring and controlling.

Fixed-time cooperative control is currently a hot research topic in multi-agent systems since it can provide a guaranteed settling time, which does not depend on initial conditions. Compared with asymptotic cooperative control algorithms, fixed-time cooperative control algorithms can provide better closed-loop performance and disturbance rejection properties. Different from finite-time control, fixed-time cooperative control can produce a faster convergence rate and provide an explicit estimate of the settling time independent of initial conditions, which is desirable for multi-agent systems. This monograph presents a systematic methodology for fixed-time cooperative control of multi-agent systems. Some fundamental concepts of fixed-time stability and stabilization are integrated with insight understanding. Within the framework of fixed-time stabilization, fixed-time consensus tracking problems of multi-agent systems with various dynamics are investigated and elaborated in detail. An application of fixed-time consensus for a distributed optimization is introduced and some practical scenarios are provided to show the superiority of the fixed-time methodology.

Structure and readership. This monograph is concerned with fixed-time cooperative control of multi-agent systems. in Chap. 1, a brief introduction of some developments of fixed-time cooperative control is first provided. Then a motivation of fixed-time stability and some related emergent research issues in cooperative control community for multi-agent systems are presented.

Fixed-time stability and stabilization: In Chap. 2, some fundamental concepts of finite-time stability and fixed-time stability are first introduced. Then some theory of fixed-time stability is presented and a set of sufficient conditions are established for generic nonlinear systems and explicit expressions for deriving the settling time. Based on the theory, two important methodologies are provided to solve a fixed-time stabilization control problem for certain systems.

Fixed-time consensus tracking for integrator-type multi-agent systems: In Chaps. 3, 4, and 5, fixed-time consensus tracking control problems are solved for first-order, second-order, and high-order integrator-type multi-agent systems, respectively. In Chap. 3, a basic idea of fixed-time consensus is illustrated by considering the simplest single integrator networks. The idea is then extended to second-order networks in Chap. 4. Since a singularity issue occurs when applying recursively the terminal sliding mode design, a nonsingular fixed-time consensus protocol that can achieve fixed-time convergence is introduced. Different from a terminal sliding mode design in Chap. 4, a fixed-time observer-based homogeneous consensus tracking protocol is presented in Chap. 5 for high-order integrator networks, which is a non-recursive singularity-free design.

Fixed-time consensus tracking for nonholonomic chained-form multi-agent systems: In Chap. 6, a fixed-time consensus tracking control problem is solved for nonholonomic chained-form multi-agent systems. A distributed observer is first proposed for each follower to estimate the leader state and the leader input in fixed time. Then, based on the observer and by adding a power integrator, a nonlinear protocol is designed such that the estimated leader state is tracked in fixed time.

Distributed optimization of multi-agent systems: In Chaps. 7 and 8, distributed optimization for multi-agent systems is investigated by using a fixed-time consensus approach. In Chap. 7, distributed protocols are proposed for both cases of time-invariant and time-varying cost functions to achieve the state agreement in a fixed time while the sum of local convex functions known to individual agents is minimized. In Chap. 8, the preservation of network connectivity is further taken into account when solving a distributed optimization problem.

Acknowledgements. We would like to acknowledge (i) Dr. Xian-Ming Zhang, Dr. Xiaohua Ge, and Dr. Lei Ding for their constructive feedback on this monograph; (ii) collaborations with Prof. Zhengtao Ding, A/Prof. Michael Defoort, Prof. Bailing Tian, and A/Prof. Lin Tie on the work of consensus tracking of first-order and high-order multi-agent systems in this monograph; (iii) supports from the Australian Research Council Discovery Project under Grant DP160103567; and the National Natural Science Foundation of China under Grants 61203022 and 61673034; and (iv) Mr. Anthony Doyle, Executive Editor, Engineering, Springer, 236 Gray's Inn Road, Floor 6, London WC1X 8HL, UK, for his encouragement to write this monograph.

Beijing, China — Zongyu Zuo
Melbourne, Australia — Qing-Long Han
Melbourne, Australia — Boda Ning

Contents

Acronyms

A^{-1}	Inverse of an invertible matrix A
A^T	Transpose of a matrix A
$A \geq 0$	Symmetric positive semi-definite
$A > 0$	Symmetric positive definite
$A \leq 0$	Symmetric negative semi-definite
$A < 0$	Symmetric negative definite
$\lambda_{\max}(A)$	Maximum eigenvalue of a real symmetric matrix A
$\lambda_{\min}(A)$	Minimum eigenvalue of a real symmetric matrix A
$\lambda_2(\cdot)$	The second smallest eigenvalue of a real symmetric matrix
$\|x\|$	Absolute value (or modulus) of x
$\|\|x\|\|$	2-norm of a real vector x
$\|\|x\|\|_p$	p-norm of a real vector x
$\text{diag}(a_1, \ldots, a_p)$	A diagonal matrix with diagonal entries a_1 to a_p
$\mathbb{R}$	Set of real numbers
$\mathbb{R}^+$	Set of positive real numbers
$\mathbb{R}^m$	m-dimensional real Euclidean space
$\mathbb{R}^{N \times m}$	Space of $N \times m$ real matrices
sign	Signum function
$\text{tr}(A)$	Trace of matrix A
$0_{N \times m}$	Zero matrix of dimension $N \times m$
$\mathscr{K}(\cdot)$	Differential inclusion
$D^* f$	Upper right-hand derivative of a function $f(t)$
$\mathbf{1}_N$	Column vector of all ones with N entries
$\mathbf{0}$	Column vector of all zeros
I_m	$m \times m$ identity matrix
C^1	A class of continuously differentiable functions
lim	Limit

sup	Supremum
inf	Infimum
$\lceil x \rfloor^{\alpha}$	$= \lvert x\rvert^{\alpha}\mathrm{sign}(x)$ for $\forall x \in \mathbb{R}$ and $\forall \alpha \in \mathbb{R}^{+}$
$\lceil x \rfloor^{\alpha}$	$= [\lvert x_1\rvert^{\alpha}\mathrm{sign}(x_1), \lvert x_2\rvert^{\alpha}\mathrm{sign}(x_2), \ldots, \lvert x_N\rvert^{\alpha}\mathrm{sign}(x_N)]$ for $\forall x \in \mathbb{R}^{N}$

Chapter 1
Introduction

In this chapter, we first introduce some underlying concepts of cooperative control. Then we briefly review how fixed-time cooperative control is motivated and promoted in the control community. Moreover, we present some related work of distributed optimization and network connectivity. Furthermore, we summarize future research topics. Finally, we give a brief outline of the monograph.

1.1 Multi-Agent Systems

A multi-agent system is a system consisting of a group of self-organizing agents with local interaction, and its macro-behavior formed spontaneously on the whole is called collective behavior. The conception of a multi-agent system, which is also known as a swarm system, stemmed from a variety of biological and social phenomena, such as schooling of fish in avoidance of harm in Fig. 1.1 [1], migration of geese in formation in Fig. 1.2 [2], cooperation of ant colony in Fig. 1.3 [3], swarming of sheep in Fig. 1.4 [4], and so on. From these phenomena, one can see that time efficiency and energy efficiency can be gained from collective behavior. Since the 1990s, American Santa Fe Institute has spurred a great worldwide interest in research on multi-agent systems [5, 6].

From the perspective of system dynamics, a simple discrete-time model is proposed [7] to study self-ordered motions of n autonomous agents moving in the plane with the same speed but with different headings. Each agent updates its heading using the nearest neighbor rule, i.e., an average of its own heading plus the headings of its neighbors. This work presents various compelling simulation results and discovers that all agents eventually move in the same direction and reach a consensus phenomenon albeit the absence of centralized coordination. The Vicsek model is, in fact, a simplified version of a model that mimics flocking and schooling behaviors in nature. A formal theoretical explanation for observed behavior in [7] is given in [8],

Z. Zuo et al., *Fixed-Time Cooperative Control of Multi-Agent Systems*,
https://doi.org/10.1007/978-3-030-20279-8_1

Fig. 1.1 Biological behavior: schooling of fish [1]

Fig. 1.2 Biological behavior: migration of geese [2]

where graph theory and nonnegative matrix theory are exploited in theoretical study of emergent collective behavior.

From the perspective of system control, an agent may refer to a sensor, a robot, an air vehicle or even a computer program, and so on. Compared with autonomous vehicles that perform solo missions, greater efficiency and operational capability can be achieved from teams of autonomous vehicles working in a cooperative fashion [9].

Fig. 1.3 Biological behavior: cooperation of ant colony [3]

Fig. 1.4 Biological behavior: swarming of sheep [4]

There are a wide range of applications of multi-agent systems, such as space-based interferometers, surveillance, combat, reconnaissance systems, source seeking and exploration including odor source localization [10, 11], and distributed reconfigurable sensor networks [12–15]. However, various cooperative controllers need to be developed to enable these applications [16–19].

During the past two decades in the control community, research on cooperative control including consensus, flocking, swarm, rendezvous has gained considerable attention [20–24], see books [25–27] and references therein. The key objective of cooperative control is to reach a desired global group behavior by using local information shared among neighboring agents in a distributed fashion [28–30]. The cooperative control promises several advantages in implementing cooperative group tasks, such as strong adaptivity, high robustness, flexibility, scalability, and low operational costs [31–33]. In what follows, we introduce some key issues in cooperative control.

- *Consensus* is a fundamental issue of cooperative control of multi-agent systems. The consensus control is to find protocols (or control laws), which enable all agents to reach an agreement on certain quantities of interest based on local relative information shared with neighbors. Consensus is a prerequisite to achieve coordination and cooperation for multi-agent systems [34–38]. For example, a consensus algorithm can be applied for formation control by appropriately choosing the information states on which consensus is reached [34].
- *Flocking* is a form of collective behavior of a large number of interacting agents with a common group goal [39]. Flocking was first investigated by Renolds in 1987 for simulating visually an aggregate motion of a flock of birds, a herd of land animals, or a school of fish for the animation industry [40]. Three heuristic rules are introduced in [40] to create flocking: cohesion, separation, and alignment, which can be viewed as a mechanism for achieving velocity synchronization and regulation of inter-agent distances based on local interaction rules [41].
- *Swarm* is a physical realization of self-organizing distributed multi-agent systems. It is well known that swarming behavior has certain advantages including avoiding predators and increasing a chance of finding food in nature. These cooperative principles from such living beings can be used in engineering for developing distributed cooperative control for autonomous agent systems such as autonomous unmanned undersea, land, or air vehicles [42]. An advantage of swarm models is the increased security since information from a subset of swarm does not reveal the final configuration [43].
- *Rendezvous* means an agreement over the location of agents in the network. Cooperative strike or cooperative jamming are two examples of a rendezvous problem [44]. From the control point of view, a multi-agent rendezvous problem is to design local control strategies, one for each agent, that enable all agents in a group to eventually rendezvous at a single unspecified location without any active communication between agents [45].

1.2 Finite-Time Cooperative Control

The settling time, which characterizes the rate of convergence of a closed-loop system, is well recognized as one of the performance specifications for control system design. Fast convergence is usually pursued in practice in order to achieve better performance. In the context of cooperative control, a pioneering result [46] shows that the second smallest eigenvalue of the Laplacian matrix of an interaction graph, i.e., *algebraic connectivity*, determines the consensus convergence rate. To increase the convergence rate, some investigations are carried out by properly designing an interaction graph among agents [47]. Through working out an optimum vertex positional configuration in the presence of a proximity constraint, consensus speed is accelerated.

In practical multi-agent systems, it is of particular interest to realize cooperative control in a finite time to meet specific system requirements. Therefore, *finite-time cooperative control*, particularly finite-time consensus [48–50], has drawn considerable research attention and appreciation [51, 52]. For example, in [53], a discontinuous controller together with a pinning control scheme is proposed, where only sign information of agents' relative states is necessary to achieve finite-time consensus. In [54], a switching strategy combining existing continuous and discontinuous finite-time consensus protocols is developed to obtain a fast convergence rate. In [55], a containment control protocol is proposed for nonlinear multi-agent systems subject to external disturbances based on a sliding mode technique. In [56], distributed coordinated tracking algorithms are proposed for agents with single-integrator dynamics and double-integrator dynamics using a variable structure approach. In [57], an algorithm based on terminal sliding mode control is proposed to realize robust finite-time consensus tracking under fixed topology. For switching topology, finite-time tracking is achieved by a pinning control technique based on LaSalle's invariance principle [58]. Since mobile agents may be governed by nonlinear inherent dynamics, finite-time leader-following consensus has been further explored for systems subjected to discontinuous inherent dynamics [59].

1.3 Fixed-Time Cooperative Control

For finite-time cooperative control of multi-agent systems, although convergence may be pursued in a finite time, an estimate of the settling time relies explicitly on initial conditions of systems in the group. This may limit the application scope of those existing results in finite-time cooperative control when information of initial states of multi-agent systems is unknown or unavailable *a priori*. Fortunately, a strategy, i.e., *fixed-time cooperative control*, has been emerging since fixed-time stability was first investigated by Polyakov in 2012 [60]. It is shown that fixed-time stability provides the guaranteed convergence (settling) time irrelevant to initial conditions. As a result, research on fixed-time cooperative control of multi-agent systems has come to the fore. In [61–63], an upper bound of the finite settling time to reach an agreement is prescribed for networked multi-agent systems with simple dynamics. A distinctive feature of this approach is that the bound is merely dependent on design parameters and algebraic connectivity of graph Laplacian, and therefore, the convergence rate of a closed-loop system may be predefined or prescribed off-line. Based on a basic idea on fixed-time consensus, some related cooperative control problems, such as bipartite consensus [64], distributed optimization [65], average tracking [66], containment control [67], dispersion and flocking [68], spacecraft rendezvous [69], are also studied. In [64], a fixed-time protocol is designed to guarantee that the states of all agents can reach an agreement in a fixed time regarding consensus values that are the same in modulus but may not be the same in sign. In [65], the problem of distributed optimization is solved by using an edge-based fixed-time consensus approach. Both time-invariant and time-varying cost functions, distributed protocols are proposed to achieve the state agreement in a fixed time while the sum of local convex functions known to individual agents is minimized. In [66], a distributed nonlinear controller is designed to ensure that a team of agents track an average of multiple time-varying reference signals under communication constraints. In [67], a distributed controller is proposed in the presence of multiple leaders such that states of followers converge to a dynamic convex hull spanned by those of leaders in a fixed time. In [68], a fixed-time consensus-based energy function is developed to guarantee system stability. An upper bound of settling time for energy consensus is obtained, and a uniform time interval is accordingly set so that energy distribution is conducted in a fair manner. In [69], a fixed-time fault-tolerant control is studied for spacecraft rendezvous and docking with a freely tumbling target.

1.3.1 Integrator-Type Multi-Agent Systems

The early work [62, 70] first independently developed fixed-time consensus protocols for single-integrator multi-agent systems, which can provide any predesignated convergence time through the appropriate selection of control gains. This appealing property can guarantee the cooperation time even for variations of the operation domain and is especially useful when initial conditions are not available in advance. The framework of fixed-time consensus design is then extended to a robust consensus problem [63] and signed networks [71] for single-integrator multi-agent systems. However, a direct extension to second-order (double-integrator) or higher order networks has come across a singularity issue [72]. This singularity issue is successfully avoided in [73, 74] by introducing a sinusoidal function into the protocol. It is worth mentioning that a fixed-time output feedback consensus protocol is also developed in [75] for second-order multi-agent systems without velocity measurements. For higher order multi-agent networks, a fixed-time consensus tracking protocol structure is developed in [76, 77], which is a non-recursive and singularity-free design method based on the work [78].

1.3.2 Complex Multi-Agent Systems

For some mechanical systems such as underactuated ships, nonholonomic dynamics can be used to better describe their dynamics [79]. Finite-time control for nonholonomic systems is investigated in [80]. Note that the results in [79, 80] are obtained for a single nonholonomic agent. For multiple agents, in [81], a decentralized feedback control strategy is proposed to solve a rendezvous problem for nonholonomic multi-agent systems. In [82], an agreement for shifted states is reached to realize formation control. Moreover, in [83], finite-time cooperative control is investigated for nonholonomic multi-agent systems. However, these results are obtained either in an infinite time horizon [81, 82], or in a finite time depending on initial states [83]. Furthermore, most of the aforementioned work only considers nonholonomic agents with a low-order model, but many mechanical systems, such as wheeled mobile robots and the knife-edge, can be described by a nonholonomic chained-form model [84]. Note that a low-order model is regarded as a special case of chained-form models. In [85], fixed-time consensus tracking is achieved for nonholonomic chained-form multi-agent systems. An observer-based nonlinear protocol is designed in a recursive manner by adding a power integrator such that the estimated leader state is tracked in a fixed time.

For other complex dynamical networked systems, some fixed-time cooperative control problems are also addressed. For example, in [86], robust fixed-time consensus tracking is investigated for double-integrator-type multi-agent systems with the application for fixed-time formation tracking of unicycle-type robots. The work [87] introduces a switching mechanism in the controller to avoid the singularity point and solves a fixed-time lag consensus problem with application to multiple single-link robotic manipulators. In [88], a fixed-time coordination control problem for telerobotics systems with asymmetric time-varying delays is addressed with real-time experimental results.

1.4 Distributed Optimization

A multi-agent consensus problem becomes a *distributed optimization* problem when the consensus value is required to minimize the sum of local cost functions known to individual agents. To solve this problem, a subgradient method is used in [89], where each agent is involved with a local averaging operation to obtain an estimate of the optimal solution, and then the estimate is projected onto a common constraint for a state update. An extension to a dual problem with inequality and equality constraints is investigated in [90]. The results in [89, 90] are obtained based on discrete-time algorithms. On the other hand, some recent work is developed for multi-agent systems with continuous-time dynamics [91–95]. For example, in [91], a distributed optimization problem is solved for continuous-time dynamics with time-varying interconnection topologies by converting the original problem to an intersection computation problem. In [95], a proportional-integral protocol is proposed to deal with inequality, equality, and bounded constraints. In [94], a distributed optimization problem is further explored with a common state set constraint, and three terms of local averaging, local projection, and local subgradient are incorporated into a controller. In [96, 97], a neurodynamic strategy is developed and a time-delay is taken into account [98] when achieving distributed optimization.

In practice, it is desirable to realize consensus in a finite time to meet specific system requirements such as tracking the position and velocity of a virtual leader in a finite time [51, 99]. Note that the aforementioned distributed optimization work [89–95] can only achieve consensus over an infinite time horizon. One exception in the literature is [100], where the finite-time convergence is taken into account for time-invariant cost functions. The bound of convergence time derived in [100] depends on initial states of agents, which may be unavailable or inaccurate in practical scenarios.

1.5 Network Connectivity

Network connectivity is the prerequisite for high-quality information exchange among agents to realize cooperative control. To better illustrate network connectivity, consider a scenario that five agents move into an unknown environment. The aim of agents is to cooperatively achieve consensus by interacting with each other. Initially, the five agents are in a configuration as shown in Fig. 1.5, where an arrow indicates that a communication link exists for interaction. For most cooperative control work, it is simply assumed that these links are preserved by default during an interaction. However, due to unknown environments or obstacles, agent "1" loses the connection with agent "2" at a particular time, as shown in Fig. 1.6. Then the connected network is broken into two separate groups, i.e., agent "1" will no longer have information exchange with the others. As a result, the agents may not reach the state agreement since agent "1" is not interacting with the others. Motivated by this scenario, a natural question is how to design a suitable protocol such that network connectivity is preserved over time while achieving cooperative tasks.

Fig. 1.5 A connected network of five agents

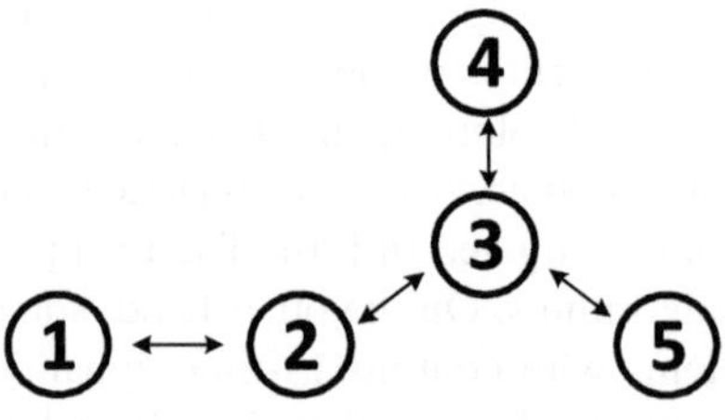

Fig. 1.6 A broken network of five agents

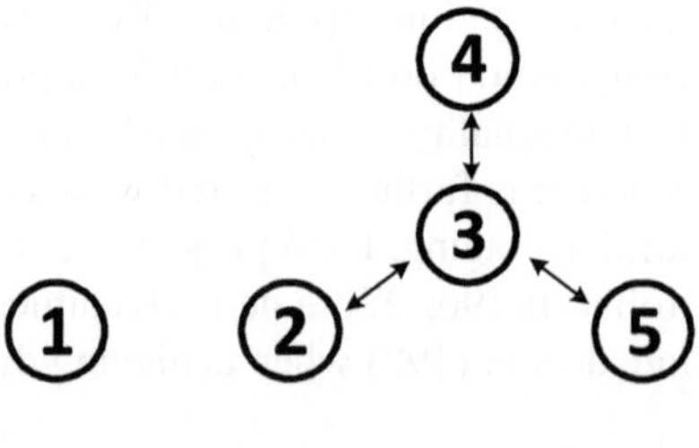

There are two major approaches to preserve the network connectivity, i.e., the *local* connectivity preservation and the *global* connectivity preservation. The former approach aims at preserving all initial connected links. The connectivity will be guaranteed if each local connection is maintained over time [101–103]. The later approach allows the deletion of redundant links, while allowing the addition of some necessary links [104, 105]. In this case, the connectivity preservation is conducted in a global manner as long as the overall interaction graph is connected. As a result, the flexibility of deletion and addition for links provides a robust and efficient communication network.

1.6 Future Research Topics

This monograph reports recent advances in fixed-time cooperative control of multi-agent systems. There still exist several issues that need to be addressed in the near future. Some important and yet challenging issues deserving research are highlighted as follows.

- *Fixed-time cooperative control with network-induced constraints*. Most of existing fixed-time control results are obtained under an assumption that communication links between controllers and actuators (or, between sensors and controllers) are perfect. However, data packet dropouts, data packet disorders, network-induced delays, and/or quantization errors are inevitable in practical networked systems [21, 29, 106, 107], see books [108, 109] and references therein. Under these networked-induced constraints, how to design proper networked fixed-time cooperative control protocols for multi-agent systems is an important and yet challenging issue;
- *Event-triggered fixed-time cooperative control*. Note that the majority of existing fixed-time cooperative control algorithms are based on a periodic sampling or a consecutive transmission paradigm, which means that each agent's state should be sampled regularly at equidistant sampling time irrespective of the current state of the system and resource usage or should be transmitted continually over a communication network at every instant of time. Event-triggered mechanisms have now shown their strengths in reducing the data sampling/transmission frequency while preserving desired system performance [110–115]. It should be also pointed out that in the context of fixed-time cooperative control, one should carefully design an event-triggered mechanism over the fixed-time interval, i.e., the time interval from the initial time to the exact settling time, such that Zeno phenomenon can be effectively excluded after the computed exact settling time. Thus, how to well address the problem of fixed-time cooperative control under event-triggered sampling/transmission deserves a deep investigation;
- *Prescribed-time or specified-time cooperative control*. The fixed-time stabilization control design ensures a uniformly bounded settling time in terms of a constant, which certainly leads to a quite conservative estimate of the settling time due

to global stability. In [116], time-varying feedback control is proposed to stabilize normal-form nonlinear systems in a prescribed finite time, which is, in fact, optimal control with a terminal constraint by employing time-varying gains. In [117], based on Pontryagin's maximum principle on Lie group, formation tracking control for vehicles is achieved with a prescribed time. In [118], from the motion-planning perspective, a class of sample-data based protocols are developed to solve a prescribed-time consensus-related control problem for second-order multi-agent systems over directed and periodically switching topologies. However, prescribed-time or specified-time cooperative control for nonlinear multi-agent systems, especially with complex dynamics, is still an open problem;

- *Fixed-time cooperative estimation and control under cyber attacks.* Cyber security has become an essential and practical issue with the fast development of real-world network systems [119–122]. As a fundamental issue in control, filtering/estimation aims at developing an algorithm to provide an estimate of an unavailable state signal of the system through the disturbed plant and/or noisy measurement [123, 124]. However, how to protect networked multi-agent systems without sacrificing the state convergence performance under cyber attacks deserves a deep investigation. In particular, how to develop secure fixed-time cooperative filtering/estimation algorithms and how to achieve fixed-time cooperative control under cyber attacks are promising future directions; and
- *Discretization of fixed-time stable cooperative systems.* For implementation in networked environments, fixed-time protocols have to be calculated in a digital controller with discrete measurements. Unfortunately, it is shown in the existing literature [125, 126] that a discrete-time system with a power rule has limit cycles rather than an origin or even exhibits unstable behaviors under improper discretization schemes, which is quite different from its continuous counterpart, although the design or analysis is performed in continuous-time scenarios. In other words, a fixed-time convergence property in discrete-time realization is impossible. However, fixed-time stable systems with a sampled continuous-time control input can still preserve some appealing properties like higher accuracy and faster convergence if a discretization scheme is carefully chosen or designed, such as an implicit Euler scheme in [127] for homogeneous systems which guarantees asymptotic convergence of the approximating solutions. Hence, discretization schemes for fixed-time stable cooperative control systems with an appropriate sampling mechanism that preserve performance of the continuous-time counterpart except for the fixed-time convergence deserve further attention and research.

1.7 Book Outline

In this monograph, we focus on fixed-time cooperative control of multi-agent systems with various agent dynamics and applications of fixed-time to distributed optimization of multi-agent systems. The monograph is organized as follows. Chapter 1 briefly reviews recent developments of fixed-time cooperative control and some related control topics. Chapter 2 provides basic definitions of fixed-time stability and recalls the results of fixed-time stability and stabilization, which are used throughout the monograph. Chapter 3 considers a fixed-time consensus tracking problem for simple integrator-type multi-agent systems, which shows a fundamental framework for fixed-time protocol design. In Chap. 4, a terminal sliding mode tool is applied to solve a fixed-time consensus tracking problem for double integrator-type multi-agent systems. In Chap. 5, based on a homogeneity tool, a fixed-time consensus tracking design method is further extended to high-order integrator-type multi-agent systems. Chapter 6 deals with a consensus tracking problem of a multi-agent system with non-holonomic chained-form dynamics. Chapters 7 and 8 are devoted to applications of a fixed-time consensus approach for distributed optimization of multi-agent systems. Especially in Chap. 8, the network connectivity preservation is taken into account in distributed optimization of multi-agent systems.

References

1. B. Blanknov, Friday smorgasbord: school is out version, available at https://reefbuilders.com/2016/11/11/friday-smorgasbord-58/. Accessed 19 Mar 2019
2. Lee's Birdwatching adventures plus, mixed flock of birds flying in a V formation, available at https://leesbird.com/2013/07/05/. Accessed 19 Mar 2019
3. Renaisscience, Mill of humanity: swarms, emergence and us, available at https://medium.com/@renaisscience/mill-of-humanity-swarms-emergence-and-us-b8f9ef50a11e/. Accessed 19 Mar 2019
4. L. Felzmann, How the science of swarms can help us fight cancer and predict the future, available at https://www.wired.com/2013/03/powers-of-swarms/. Accessed 19 Mar 2019
5. M.M. Waldrop, *Complexity: The Emerging Science at the Edge of Order and Chaos* (Simon & Schuster, New York, 1992)
6. W.B. Arthur, Complexity and the economy. Science **284**(5411), 107–109 (1999)
7. T. Vicsek, A. Czirók, E. Ben-Jacob, I. Cohen, O. Shochet, Novel type of phase transitions in a system of self-driven particles. Phys. Rev. Lett. **75**(6), 1226–1229 (1995)
8. A. Jadbabaie, J. Lin, A.S. Morse, Coordination of groups of mobile autonomous agents using nearest neighbor rules. IEEE Trans. Autom. Control **48**(6), 988–1001 (2003)
9. W. Ren, R.W. Beard, E.M. Atkins, Information consensus in multivehicle cooperative control. IEEE Control Syst. Mag. **27**(2), 71–82 (2007)
10. Q. Lu, Q.-L. Han, X. Xie, S. Liu, A finite-time motion control strategy for odor source localization. IEEE Trans. Ind. Electron. **61**(10), 5419–5430 (2014)
11. Q. Lu, Q.-L. Han, S. Liu, A finite-time particle swarm optimization algorithm for odor source localization. Inf. Sci. **277**, 111–140 (2014)
12. Y. Cao, W. Yu, W. Ren, G. Chen, An overview of recent progress in the study of distributed multi-agent coordination. IEEE Trans. Ind. Inf. **9**(1), 427–438 (2013)
13. L. Ding, Q.-L. Han, G. Guo, Network-based leader-following consensus for distributed multi-agent systems. Automatica **49**(7), 2281–2286 (2013)
14. X. Ge, Q.-L. Han, Consensus of multiagent systems subject to partially accessible and overlapping Markovian network topologies. IEEE Trans. Cybern. **47**(8), 1807–1819 (2017)
15. X. Ge, Q.-L. Han, F. Yang, Event-based set-membership leader-following consensus of networked multi-agent systems subject to limited communication resources and unknown-but-bounded noise. IEEE Trans. Ind. Electron. **64**(6), 5045–5054 (2017)

16. Y. Cao, W. Ren, Z. Meng, Decentralized finite-time sliding mode estimators and their applications in decentralized finite-time formation tracking. Syst. Control Lett. **59**(9), 522–529 (2010)
17. X. Ge, Q.-L. Han, Distributed formation control of networked multi-agent systems using a dynamic event-triggered communication mechanism. IEEE Trans. Ind. Electron. **64**(10), 8118–8127 (2017)
18. C. Wang, Z. Zuo, Z. Lin, Z. Ding, Consensus control of a class of Lipschitz nonlinear systems with input delay. IEEE Trans. Circuits Syst. I: Regul. Papers **62**(11), 2730–2738 (2015)
19. X. Ge, Q.-L. Han, D. Ding, X.-M. Zhang, B. Ning, A survey on recent advances in distributed sampled-data cooperative control of multi-agent systems. Neurocomputing **275**, 1684–1701 (2018)
20. W. He, C. Xu, Q.-L. Han, F. Qian, Z. Lang, Finite-time L_2 leader-follower consensus of networked Euler-Lagrange systems with external disturbances. IEEE Trans. Syst. Man Cybern.: Syst. **48**(11), 1920–1928 (2018)
21. X.-M. Zhang, Q.-L. Han, X. Yu, Survey on recent advances in networked control systems. IEEE Trans. Ind. Inf. **12**(5), 1740–1752 (2016)
22. C. Wang, Z. Zuo, Z. Lin, Z. Ding, A truncated prediction approach to consensus control of Lipschitz nonlinear multiagent systems with input delay. IEEE Trans. Control Netw. Syst. **4**(4), 716–724 (2017)
23. Q. Lu, Q.-L. Han, S. Liu, A cooperative control framework for collective decision on movement behaviors of particles. IEEE Trans. Evol. Comput. **20**(6), 859–873 (2016)
24. Q. Lu, Q.-L. Han, Mobile robot networks for environmental monitoring: a cooperative receding horizon temporal logic control approach. IEEE Trans. Cybern. **49**(2), 698–711 (2019)
25. F.L. Lewis, H. Zhang, K. Hengster-Movric, A. Das, *Cooperative Control of Multi-Agent Systems: Optimal and Adaptive Design Approaches* (Springer, London, 2014)
26. Z. Li, Z. Duan, *Cooperative Control of Multi-Agent Systems: A Consensus Region Approach* (Taylor & Francis Inc, United States, 2014)
27. Y. Song, Y. Wang, *Cooperative Control of Nonlinear Networked Systems: Infinite-time and Finite-time Design Methods* (Springer International Publishing, 2019)
28. L. Ding, Q.-L. Han, L.Y. Wang, E. Sindi, Distributed cooperative optimal control of DC microgrids with communication delays. IEEE Trans. Ind. Inf. **14**(9), 3924–3935 (2018)
29. X. Ge, F. Yang, Q.-L. Han, Distributed networked control systems: a brief overview. Inf. Sci. **380**, 117–131 (2017)
30. W. He, G. Chen, Q.-L. Han, W. Du, J. Cao, F. Qian, Multiagent systems on multilayer networks: synchronization analysis and network design. IEEE Trans. Syst. Man Cybern. Syst. **47**(7), 1655–1667 (2017)
31. B. Tian, L. Yin, H. Wang, Finite-time reentry attitude control based on adaptive multivariable disturbance compensation. IEEE Trans. Ind. Electron. **62**(9), 5889–5898 (2015)
32. T. Zhao, Z. Ding, Distributed finite-time optimal resource management for microgrids based on multi-agent framework. IEEE Trans. Ind. Electron. **65**(8), 6571–6580 (2018)
33. X.-M. Zhang, Q.-L. Han, B.-L. Zhang, An overview and deep investigation on sampled-data-based event-triggered control and filtering for networked systems. IEEE Trans. Ind. Inf. **13**(1), 4–16 (2017)
34. W. Ren, Consensus based formation control strategies for multi-vehicle systems, In *Proceedings of the American Control Conference (ACC)* (Minnesota, USA, Minneapolis, 2006), pp. 4237–4242
35. W. He, B. Zhang, Q.-L. Han, F. Qian, J. Kurths, J. Cao, Leader-following consensus of nonlinear multiagent systems with stochastic sampling. IEEE Trans. Cybern. **47**(2), 327–338 (2017)
36. X. Ge, Q.-L. Han, X.-M. Zhang, Achieving cluster formation of multi-agent systems under aperiodic sampling and communication delays. IEEE Trans. Ind. Electron. **65**(4), 3417–3426 (2018)
37. Y.-L. Wang, Q.-L. Han, Network-based fault detection filter and controller coordinated design for unmanned surface vehicles in network environments. IEEE Trans. Ind. Inf. **12**(5), 1753–1765 (2016)

38. L. Ding, Q.-L. Han, X. Ge, X.-M. Zhang, An overview of recent advances in event-triggered consensus of multiagent systems. IEEE Trans. Cybern. **48**(4), 1110–1123 (2018)
39. R. Olfati-Saber, Flocking for multi-agent dynamic systems: algorithms and theory. IEEE Trans. Autom. Control **51**(3), 401–420 (2006)
40. C.W. Reynolds, Flocks, herds, and schools: a distributed behavioral model. Comput. Gr. **21**(4), 25–34 (1987)
41. H.G. Tanner, A. Jadbabaie, G.J. Pappas, Flocking in fixed and switching networks. IEEE Trans. Autom. Control **52**(5), 863–868 (2007)
42. V. Gazi, K.M. Passino, Stability analysis of swarms. IEEE Trans. Autom. Control **48**(4), 692–697 (2003)
43. G. Beni, P. Liang, Pattern reconfiguration in swarms—convergence of a distributed asynchronous and bounded iterative algorithm. IEEE Trans. Robot. Autom. **12**(3), 485–490 (1996)
44. A. Tiwari, J. Fung, J.M. Carson, A framework for Lyapunov certificates for multi-vehicle rendezvous problem, In *Proceedings of the American Control Conference (ACC)* (Massachusetts, USA, Boston, 2004), pp. 5582–5587
45. J. Lin, A.S. Morse, B.D.O. Anderson, The multi-agent rendezvous problem—part 1: the synchronous case. SIAM J. Control Optim. **46**(6), 2096–2119 (2007)
46. R. Olfati-Saber, R.M. Murray, Consensus problems in networks of agents with switching topology and time-delays. IEEE Trans. Autom. Control **49**(9), 1520–1533 (2004)
47. Y. Kim, M. Mesbahi, On maximizing the second smallest eigenvalue of a state-dependent graph Laplacian. IEEE Trans. Autom. Control **51**(1), 116–120 (2006)
48. L. Wang, F. Xiao, Finite-time consensus problems for networks of dynamic agents. IEEE Trans. Autom. Control **55**(4), 950–955 (2010)
49. F. Xiao, L. Wang, J. Chen, Y. Gao, Finite-time formation control for multi-agent systems. Automatica **45**(11), 2605–2611 (2009)
50. Z.-H. Guan, F.-L. Sun, Y.-W. Wang, T. Li, Finite-time consensus for leader-following second-order multi-agent networks. IEEE Trans. Circuits Syst. I: Regul. Papers **59**(11) (2012)
51. Q. Lu, Q.-L. Han, B. Zhang, D. Liu, S. Liu, Cooperative control of mobile sensor networks for environmental monitoring: An event-triggered finite-time control scheme. IEEE Trans. Cybern. **47**(12), 4134–4147 (2017)
52. H. Liu, L. Cheng, M. Tan, Z. Hou, Y. Wang, Distributed exponential finite-time coordination of multi-agent systems: containment control and consensus. Int. J. Control **88**(2), 237–247 (2015)
53. G. Chen, F.L. Lewis, L. Xie, Finite-time distributed consensus via binary control protocols. Automatica **47**(9), 1962–1968 (2011)
54. X. Liu, J. Lam, W. Yu, G. Chen, Finite-time consensus of multiagent systems with a switching protocol. IEEE Trans. Neural Netw. Learn. Syst. **27**(4), 853–862 (2016)
55. Y. Zhang, Y. Yang, Y. Zhao, G. Wen, Distributed finite-time tracking control for nonlinear multi-agent systems subject to external disturbances. Int. J. Control **86**(1), 29–40 (2013)
56. Y. Cao, W. Ren, Distributed coordinated tracking with reduced interaction via a variable structure approach. IEEE Trans. Autom. Control **57**(1), 33–48 (2012)
57. S. Khoo, L. Xie, Z. Man, Robust finite-time consensus tracking algorithm for multirobot systems. IEEE/ASME Trans. Mechatron. **14**(2), 219–228 (2009)
58. Z.-H. Guan, F.-L. Sun, Y.-W. Wang, T. Li, Finite-time consensus for leader-following second-order multi-agent networks. IEEE Trans. Circuits Syst. I: Regul. Papers **59**(11), 2646–2654 (2012)
59. B. Ning, J. Jin, J. Zheng, Z. Man, Finite-time and fixed-time leader-following consensus for multi-agent systems with discontinuous inherent dynamics. Int. J. Control **91**(6), 1259–1270 (2018)
60. A. Polyakov, Nonlinear feedback design for fixed-time stabilization of linear control systems. IEEE Trans. Autom. Control **57**(8), 2106–2110 (2012)
61. S. Parsegov, A. Polyakov, P. Shcherbakov, Nonlinear fixed-time control protocol for uniform allocation of agents on a segment, In *Proceedings of the IEEE Conference on Decision and Control* (Hawaii, USA, 2013), pp. 7732–7737

62. Z. Zuo, L. Tie, A new class of finite-time nonlinear consensus protocols for multi-agent systems. Int. J. Control **87**(2), 363–370 (2014)
63. Z. Zuo, L. Tie, Distributed robust finite-time nonlinear consensus protocols for multi-agent systems. Int. J. Syst. Sci. **47**(6), 1366–1375 (2016)
64. D. Meng, Y. Jia, J. Du, Finite-time consensus for multiagent systems with cooperative and antagonistic interactions. IEEE Trans. Neural Netw. Learn. Syst. **27**(4), 762–770 (2016)
65. B. Ning, Q.-L. Han, Z. Zuo, Distributed optimization for multiagent systems: an edge-based fixed-time consensus approach. IEEE Trans. Cybern. **49**(1), 122–132 (2019)
66. H. Hong, W. Yu, X. Yu, G. Wen, A. Alsaedi, Fixed-time connectivity-preserving distributed average tracking for multi-agent systems. IEEE Trans. Circuits Syst. II: Express Br. **64**(10), 1192–1196 (2017)
67. B. Ning, Z. Zuo, J. Jin, J. Zheng, Distributed fixed-time coordinated tracking for nonlinear multi-agent systems under directed graphs. Asian J. Control **20**(2), 646–658 (2018)
68. B. Ning, Q.-L. Han, Z. Zuo, J. Jin, J. Zheng, Collective behaviors of mobile robots beyond the nearest neighbor rules with switching topology. IEEE Trans. Cybern. **48**(5), 1577–1590 (2018)
69. B. Jiang, Q. Hu, M.-I. Friswell, Fixed-time rendezvous control of spacecraft with a tumbling target under loss of actuator effectiveness. IEEE Trans. Aerosp. Electron. Syst. **52**(4), 1576–1586 (2016)
70. S. Parsegov, A. Polyakov, P.S. Shcherbakov, Fixed-time consensus algorithm for multi-agent systems with integrator dynamics, In *Proceedings of the IFAC Workshop on Distributed Estimation and Control in Networked Systems* (Koblenz, Germany, 2013), pp. 110–115
71. D. Meng, Z. Zuo, Signed-average consensus for networks of agents: a nonlinear fixed-time convergence protocol. Nonlinear Dyn. **85**(1), 155–165 (2016)
72. Y. Feng, X. Yu, Z. Man, On nonsingular terminal sliding-mode control of nonlinear systems. Automatica **49**(6), 1715–1722 (2013)
73. Z. Zuo, Nonsingular fixed-time consensus tracking for second-order multi-agent networks. Automatica **54**, 305–309 (2015)
74. J. Fu, J. Wang, Fixed-time coordinated tracking for second-order multi-agent systems with bounded input uncertainties. Syst. Control Lett. **93**, 1–12 (2016)
75. B. Tian, H. Lu, Z. Zuo, W. Yang, Fixed-time leader-following output feedback consensus for second-order multiagent systems. IEEE Trans. Cybern. **49**(4), 1545–1550 (2019)
76. Z. Zuo, B. Tian, M. Defoort, Z. Ding, Fixed-time consensus tracking for multi-agent systems with high-order integrator dynamics. IEEE Trans. Autom. Control **63**(2), 563–570 (2018)
77. Z. Zuo, Q.-L. Han, B. Ning, An explicit estimate for the upper bound of the settling time in fixed-time leader-following consensus of high-order multi-variable multi-agent systems. IEEE Trans. Ind. Electron. **66**(8), 6250–6259 (2019)
78. M. Basin, Y. Shtessel, F. Aldukali, Continuous finite- and fixed-time high-order regulators. J. Frankl. Inst. **253**(18), 5001–5012 (2016)
79. Z.-P. Jiang, Global tracking control of underactuated ships by Lyapunov's direct method. Automatica **38**(2), 301–309 (2002)
80. Y. Wu, B. Wang, G. Zong, Finite-time tracking controller design for nonholonomic systems with extended chained form. IEEE Trans. Circuits Syst. II: Express Br. **52**(11), 798–802 (2005)
81. D.V. Dimarogonas, K.J. Kyriakopoulos, On the rendezvous problem for multiple nonholonomic agents. IEEE Trans. Autom. Control **52**(5), 916–922 (2007)
82. W. Dong, J.A. Farrell, Cooperative control of multiple nonholonomic mobile agents. IEEE Trans. Autom. Control **53**(6), 1434–1448 (2008)
83. M. Ou, H. Du, S. Li, Finite-time formation control of multiple nonholonomic mobile robots. Int. J. Robust Nonlinear Control **24**(1), 140–165 (2014)
84. H. Du, G. Wen, X. Yu, S. Li, M.Z. Chen, Finite-time consensus of multiple nonholonomic chained-form systems based on recursive distributed observer. Automatica **62**, 236–242 (2015)

85. B. Ning, Q.-L. Han, Prescribed finite-time consensus tracking for multi-agent systems with nonholonomic chained-form dynamics. IEEE Trans. Autom. Control **64**(4), 1686–1693 (2019)
86. X. Chu, Z. Peng, G. Wen, A. Rahmani, Robust fixed-time consensus tracking with application to formation control of unicycles. IET Control Theory Appl. **12**(1), 53–59 (2018)
87. J. Ni, L. Liu, C. Liu, J. Liu, Fixed-time leader-following consensus for second-order multi-agent systems with input delay. IEEE Trans. Ind. Electron. **64**(11), 8635–8646 (2017)
88. Y. Yang, C. Hua, J. Li, X. Guan, Robust adaptive uniform exact tracking control for uncertain Euler-Lagrange system. Int. J. Control **90**(12), 2711–2720 (2017)
89. A. Nedić, A. Ozdaglar, P.A. Parrilo, Constrained consensus and optimization in multi-agent networks. IEEE Trans. Autom. Control **55**(4), 922–938 (2010)
90. D. Yuan, S. Xu, H. Zhao, Distributed primal-dual subgradient method for multiagent optimization via consensus algorithms. IEEE Trans. Syst. Man Cybern. Part B **41**(6), 1715–1724 (2011)
91. G. Shi, K.H. Johansson, Y. Hong, Reaching an optimal consensus: dynamical systems that compute intersections of convex sets. IEEE Trans. Autom. Control **58**(3), 610–622 (2013)
92. Q. Liu, J. Wang, A second-order multi-agent network for bound-constrained distributed optimization. IEEE Trans. Autom. Control **60**(12), 3310–3315 (2015)
93. S.S. Kia, J. Cortés, S. Martínez, Distributed convex optimization via continuous-time coordination algorithms with discrete-time communication. Automatica **55**, 254–264 (2015)
94. Z. Qiu, S. Liu, L. Xie, Distributed constrained optimal consensus of multi-agent systems. Automatica **68**, 209–215 (2016)
95. S. Yang, Q. Liu, J. Wang, A multi-agent system with a proportional-integral protocol for distributed constrained optimization. IEEE Trans. Autom. Control **62**(7), 3461–3467 (2017)
96. S. Yang, Q. Liu, J. Wang, A collaborative neurodynamic approach to multiple-objective distributed optimization. IEEE Trans. Neural Netw. Learn. Syst. **29**(4), 981–992 (2018)
97. Q. Liu, S. Yang, J. Wang, A collective neurodynamic approach to distributed constrained optimization. IEEE Trans. Neural Netw. Learn. Syst. **28**(8), 1747–1758 (2017)
98. S. Yang, Q. Liu, J. Wang, Distributed optimization based on a multiagent system in the presence of communication delays. IEEE Trans. Syste. Man Cybern.: Syst. **47**(5), 717–728 (2017)
99. Y. Shang, Finite-time scaled consensus through parametric linear iterations. Int. J. Syste. Sci. **48**(10), 2033–2040 (2017)
100. P. Lin, W. Ren, Y. Song, J. A. Farrell, Distributed optimization with the consideration of adaptivity and finite-time convergence, In *Proceedings of the American Control Conference (ACC)* (Oregon, USA, Portland, 2014), pp. 3177–3182
101. G. Notarstefano, K. Savla, F. Bullo, A. Jadbabaie, Maintaining limited-range connectivity among second-order agents, In *Proceedings of the American Control Conference (ACC)* (Minnesota, USA, Minneapolis, 2006), pp. 2124–2129
102. M. Ji, M. Egerstedt, Distributed coordination control of multiagent systems while preserving connectedness. IEEE Trans. Robot. **23**(4), 693–703 (2007)
103. D.V. Dimarogonas, K.J. Kyriakopoulos, Connectedness preserving distributed swarm aggregation for multiple kinematic robots. IEEE Trans. Robot. **24**(5), 1213–1223 (2008)
104. M.M. Zavlanos, G.J. Pappas, Distributed connectivity control of mobile networks. IEEE Trans. Robot. **24**(6), 1416–1428 (2008)
105. R.K. Williams, G.S. Sukhatme, Constrained interaction and coordination in proximity-limited multi-agent systems. IEEE Trans. Robot. **29**(4), 930–944 (2013)
106. Y.-L. Wang, Q.-L. Han, Network-based modelling and dynamic output feedback control for unmanned marine vehicles in network environments. Automatica **91**, 43–53 (2018)
107. Y.-L. Wang, Q.-L. Han, M.-R. Fei, C. Peng, Network-based T-S fuzzy dynamic positioning controller design for unmanned marine vehicles. IEEE Trans. Cybern. **48**(9), 2750–2763 (2018)
108. D. Yue, C. Peng, Q.-L. Han, *Analysis and Synthesis of Networked Control Systems* (Science Press, Beijing, P.R. China, 2007)

109. C. Peng, D. Yue, Q.-L. Han, *Communication and Control for Networked Complex Systems* (Springer, Berlin, 2015)
110. D. Yue, E. Tian, Q.-L. Han, A delay system method for designing event-triggered controllers of networked control systems. IEEE Trans. Autom. Control **58**(2), 475–481 (2013)
111. C. Peng, Q.-L. Han, D. Yue, To transmit or not to transmit: a discrete event-triggered communication scheme for networked Takagi-Sugeno fuzzy systems. IEEE Trans. Fuzzy Syst. **21**(1), 164–170 (2013)
112. C. Peng, Q.-L. Han, A novel event-triggered transmission scheme and L_2 control co-design for sampled-data control systems. IEEE Trans. Autom. Control **58**(10), 2620–2626 (2013)
113. C. Peng, Q.-L. Han, On designing a novel self-triggered sampling scheme for networked control systems with data losses and communication delays. IEEE Trans. Ind. Electron. **63**(2), 1239–1248 (2016)
114. X.-M. Zhang, Q.-L. Han, A decentralized event-triggered dissipative control scheme for systems with multiple sensors to sample the system outputs. IEEE Trans. Cybern. **46**(12), 2745–2757 (2016)
115. X. Ge, Q.-L. Han, Z. Wang, A dynamic event-triggered transmission scheme for distributed set-membership estimation over wireless sensor networks. IEEE Trans. Cybern. **49**(1), 171–183 (2019)
116. Y. Song, Y. Wang, J. Holloway, M. Krstic, Time-varying feedback for regulation of normal-form nonlinear systems in prescribed finite time. Automatica **83**, 243–251 (2017)
117. Y. Liu, Y. Zhao, G. Chen, Finite-time formation tracking control for multiple vehicles: a motion planning approach. Int. J. Robust Nonlinear Control **26**(14), 3130–3149 (2016)
118. Y. Liu, Y. Zhao, W. Ren, G. Chen, Appointed-time consensus: Accurate and practical designs. Automatica **89**, 425–429 (2018)
119. L. Ma, Z. Wang, Q.-L. Han, H.-K. Lam, Variance-constrained distributed filtering for time-varying systems with multiplicative noises and deception attacks over sensor networks. IEEE Sens. J. **17**(7), 2279–2288 (2017)
120. L. Hu, Z. Wang, Q.-L. Han, X. Liu, State estimation under false data injection attacks: Security analysis and system protection. Automatica **87**, 176–183 (2018)
121. D. Ding, Z. Wang, Q.-L. Han, G. Wei, Security control for discrete-time stochastic nonlinear systems subject to deception attacks. IEEE Trans. Syst. Man Cybern.: Syst. **48**(5), 779–789 (2018)
122. D. Ding, Q.-L. Han, Y. Xiang, X. Ge, X.-M. Zhang, A survey on security control and attack detection for industrial cyber-physical systems. Neurocomputing **275**, 1674–1683 (2018)
123. S. Zhu, Q.-L. Han, C. Zhang, L_1-gain performance analysis and positive filter design for positive discrete-time markov jump linear systems: a linear programming approach. Automatica **50**(8), 2098–2107 (2014)
124. X. Ge, Q.-L. Han, Distributed event-triggered H_∞ filtering over sensor networks with communication delays. Inf. Sci. **291**, 128–142 (2015)
125. X. Yu, J. Xu, Y. Hong, S. Yu, Analysis of a class of discrete-time systems with power rule. Automatica **43**(3), 562–566 (2007)
126. H. Du, X. Yu, M. Chen, S. Li, Chattering-free discrete-time sliding mode control. Automatica **68**, 87–91 (2016)
127. D. Efimov, A. Polyakov, A. Levant, W. Perruquetti, Realization and discretization of asymptotically stable homogeneous systems. IEEE Trans. Autom. Control **62**(11), 5962–5969 (2017)

Chapter 2
Fixed-Time Stability and Stabilization

This chapter introduces some basic results of finite-time and fixed-time stability and stabilization. To illustrate these results, three interesting examples are first presented to give an insight into the key feature on a standard structure of finite-time and fixed-time stabilization. Then formal definitions of finite-time and fixed-time stability are stated and corresponding stability theorems are established. Based on these concepts and stability theorems, two types of fixed-time stabilization control approaches, i.e., a terminal sliding mode design approach and a homogeneity and Lyapunov design approach, are provided. Note, however, that these two design approaches are only applicable for certain class of systems.

2.1 Basic Concepts and Definitions

2.1.1 Interesting Examples

Before stating definitions of finite-time and fixed-time stability, we give three examples to reveal a fundamental concept and shed light on a motivation of developing fixed-time stability.

Example 2.1.1 Consider a simple system

$$\dot{x}(t) = -x^{\frac{1}{3}}(t), \qquad x(0) = x_0, \tag{2.1}$$

where $x(t) \in \mathbb{R}$ is the state. The system trajectory satisfies that $x(t) = (x^{\frac{2}{3}}(0) - \frac{2}{3}t)^{\frac{3}{2}}$, which reveals that $x(t, x_0) = 0$ for $t \geq t_1^* = \frac{3}{2}x_0^{\frac{2}{3}}$. □

Note that Example 2.1.1 is essentially a finite-time stable system, where the system's solution reaches the origin at some finite moment. The key feature lies in the

Z. Zuo et al., *Fixed-Time Cooperative Control of Multi-Agent Systems*,
https://doi.org/10.1007/978-3-030-20279-8_2

fractional exponent that enables the system to achieve the finite-time convergence to zero. Although the faster rate of convergence can be achieved in the neighborhood of the origin, the rate of convergence becomes slow when the system state is far away from 1. As an improvement of the rate of convergence, another example is given as follows.

Example 2.1.2 Consider a scalar system

$$\dot{x}(t) = -x(t) - x^{\frac{1}{3}}(t), \qquad x(0) = x_0. \tag{2.2}$$

One can obtain that $x(t, x_0) = 0$ for $t \geq t_2^* = \frac{3}{2}\ln(x_0^{\frac{2}{3}} + 1)$. □

Noting that a linear term "$-x(t)$" is introduced, it is expected that a faster speed of convergence than the one in system (2.1) can be obtained because the linear term dominates the transient performance for any initial state $x_0 \gg 1$, which can also be verified by noting that $t_1^* > t_2^*$. Then a natural question is whether the system response can be even faster. The following example gives a positive answer.

Example 2.1.3 Consider a polynomial feedback system

$$\dot{x}(t) = -x^3(t) - x^{\frac{1}{3}}(t), \qquad x(0) = x_0. \tag{2.3}$$

Integrating both sides of system (2.3) yields

$$\begin{aligned}
&\frac{3}{2}\Big(\frac{\sqrt{2}}{8}\ln\frac{x^{\frac{4}{3}} + \sqrt{2}x^{\frac{2}{3}} + 1}{x^{\frac{4}{3}} - \sqrt{2}x^{\frac{2}{3}} + 1} + \frac{\sqrt{2}}{4}\arctan(\sqrt{2}x^{\frac{2}{3}} + 1) \\
&\quad + \frac{\sqrt{2}}{4}\arctan(\sqrt{2}x^{\frac{2}{3}} - 1) - C\Big) = -t,
\end{aligned}$$

where C denotes an integral constant about x_0. Thus, $x(t, x_0) = 0$ for $t \geq t_3^*$, where

$$\begin{aligned}
t_3^* = &\frac{3}{2}\Big(\frac{\sqrt{2}}{8}\ln\frac{x_0^{\frac{4}{3}} + \sqrt{2}x_0^{\frac{2}{3}} + 1}{x_0^{\frac{4}{3}} - \sqrt{2}x_0^{\frac{2}{3}} + 1} + \frac{\sqrt{2}}{4}\arctan(\sqrt{2}x_0^{\frac{2}{3}} + 1) \\
&+ \frac{\sqrt{2}}{4}\arctan(\sqrt{2}x_0^{\frac{2}{3}} - 1)\Big).
\end{aligned} \tag{2.4}$$

Moreover, it can be shown that $\lim_{x_0 \to \infty} t_3^* = \frac{3\sqrt{2}}{8}\pi$. □

Example 2.1.3 indicates that introducing two polynomial terms with one fractional exponent and the other one larger than 1 may achieve high-speed convergence and the upper bound of the reaching time is uniformly bounded by a constant that is not related to initial conditions.

2.1.2 Stability Definitions

Consider the nonlinear autonomous system described by

$$\dot{x}(t) = f(x(t)), \quad x(0) = x_0, \tag{2.5}$$

where $x \in \mathbb{R}^\ell$ is the state, $f : \mathscr{D} \to \mathbb{R}^\ell$ is an upper semicontinuous mapping in an open neighborhood $\mathscr{D}$ of the origin such that the set $f(x)$ is non-empty for any $x \in \mathscr{D}$, and $f(0) = 0$ for all $t > 0$. The solutions of system (2.5) are understood in the Filippov sense [1] if $f(x(t))$ is discontinuous. Let $x(t, x_0)$ be an arbitrary solution of the Cauchy problem of system (2.5).

Definition 2.1 ([2]) The equilibrium of system (2.5) at the origin is

- *Lyapunov stable*, if, for each $\varepsilon > 0$, there exists a $\delta = \delta(\varepsilon) > 0$ such that, if $\forall \|x_0\| < \delta$, then $\|x(t, x_0)\| < \varepsilon$ for $\forall t > 0$;
- *locally asymptotically stable* if it is stable and δ can be chosen such that if $\forall \|x_0\| < \delta$, then $\lim_{t\to\infty} \|x_0\| = 0$;
- *globally asymptotically stable* if it is stable and for all $x_0 \in \mathbb{R}^\ell$, $\lim_{t\to\infty} \|x_0\| = 0$;
- *unstable* if it is not stable.

Definition 2.2 ([3]) The origin is a *finite-time stable* equilibrium of system (2.5), if and only if the origin is Lyapunov stable and there exist an open neighborhood $\mathscr{N} \subseteq \mathscr{D}$ of the origin and a positive definite function $T(x_0) = \sup_{x(t,x_0)} \inf\{T \geq 0 : x(t, x_0) = 0 \text{ for all } t \geq T,\ x_0 \in \mathscr{N}\}$ called the settling time function such that, for all $x(0) \in \mathscr{N} \backslash \{0\}$, $T(x_0) < \infty$. Furthermore, the origin is a *globally finite-time stable* equilibrium if the origin is finite-time stable with $\mathscr{N} = \mathbb{R}^\ell$. Also, finite-time stability of the origin implies asymptotic stability of the origin.

Definition 2.3 ([4]) The origin is a *fixed-time stable* equilibrium of system (2.5) if it is globally finite-time stable and the settling time function $T(x_0)$ is bounded by a real number $T_{\max} > 0$, i.e., $T(x_0) \leq T_{\max}$, $\forall x_0 \in \mathbb{R}^\ell$.

By Definitions 2.2 and 2.3, one can see clearly that the systems in Examples 2.1.1 and 2.1.2 are globally finite-time stable, while Example 2.1.3 demonstrates a fixed-time stable system.

2.2 Fixed-Time Stability Theory

A Lyapunov's direct method is first recalled to give some sufficient conditions for asymptotical and finite-time stability.

Definition 2.4 ([2])
A function $V(x) : \mathscr{D} \to \mathbb{R}$ with $V(0) = 0$ is called

- *positive definite* if $V(x) > 0, \forall x \in \mathscr{D}\backslash\{0\}$;
- *positive semi-definite* if $V(x) \geq 0, \forall x \in \mathscr{D}\backslash\{0\}$;
- *negative definite* if $V(x) < 0, \forall x \in \mathscr{D}\backslash\{0\}$;
- *negative semi-definite* if $V(x) \leq 0, \forall x \in \mathscr{D}\backslash\{0\}$.

Theorem 2.1 ([2]) *Suppose that there exists a continuously differentiable positive definite function $V(x) : \mathscr{D} \to \mathbb{R}$ such that $V(x) > 0, \forall x \in \mathscr{D}\backslash\{0\}$. Then the origin is a locally stable equilibrium of system (2.5) if $\dot{V}(x(t)) \leq 0, \forall x \in \mathscr{D}$ and a locally asymptotically stable if $\dot{V}(x(t)) < 0, \forall x \in \mathscr{D}\backslash\{0\}$. In addition, if $\mathscr{D} = \mathbb{R}^\ell$ and $V(x)$ is radially unbounded, i.e., $\lim_{\|x\|\to\infty} V(x) = \infty$, these results hold globally.*

$V(x)$ is called a *Lyapunov function* for system (2.5) if it satisfies the properties above.

In the following theorem, a sufficient condition of finite-time stability is given.

Theorem 2.2 ([3]) *Suppose that there exists a continuously differentiable positive definite function $V(x) : \mathscr{D} \to \mathbb{R}$ such that for any real numbers $c > 0$ and $\alpha \in (0, 1)$, the following inequality holds:*

$$\dot{V}(x) + cV^\alpha(x) \leq 0, \quad x \in \mathscr{N}\backslash\{0\}. \tag{2.6}$$

Then the origin is a finite-time stable equilibrium of system (2.5) and the settling time is given by

$$T(x_0) \leq \frac{1}{c(1-\alpha)} V^{(1-\alpha)}(x_0). \tag{2.7}$$

Furthermore, if $\mathscr{N} = \mathscr{D} = \mathbb{R}^\ell$, V is radially unbounded and $\dot{V} < 0$ on $\mathbb{R}^\ell\backslash\{0\}$, the origin is a globally finite-time stable equilibrium of system (2.5).

Applying Theorem 2.2, one can derive the following corollaries [5, 6], which generalize Example 2.1.3 and present two first-order fixed-time stable systems with the uniformly bounded settling time.

Corollary 2.1 *Consider a scalar system*

$$\dot{x}(t) = -\alpha x^{2-\frac{p}{q}}(t) - \beta x^{\frac{p}{q}}(t), \quad x(0) = x_0, \tag{2.8}$$

where $\alpha, \beta > 0$, and p, q satisfying $q > p > 0$ are odd integers. The equilibrium of system (2.8) is fixed-time stable and the settling time is given by

$$T(x_0) \leq T_{\max} := \frac{q\pi}{2\sqrt{\alpha\beta}(q-p)}. \tag{2.9}$$

Proof Let $V(y) = y^2 \geq 0$. Differentiating $V(y)$ along system (2.8) yields

$$\begin{aligned}\dot{V}(y) &= 2y\left(-\alpha y^{2-\frac{p}{q}} - \beta y^{\frac{p}{q}}\right) \\ &= -2\alpha(y^2)^{\frac{3q-p}{2q}} - 2\beta(y^2)^{\frac{p+q}{2q}} \\ &= -2\left(\alpha V^{\frac{q-p}{q}} + \beta\right) V^{\frac{p+q}{2q}}.\end{aligned} \tag{2.10}$$

Note that $\alpha V^{(q-p)/q} > 0$ implies that $\dot{V}(y) \leq -2\beta V^{(p+q)/(2q)}$. In view of $0 < (p+q)/(2q) < 1$, system (2.8) is finite-time stable by Theorem 2.2. Furthermore, since $V(y) = 0$ in system (2.8) is a trivial case, assuming that $V(y) \neq 0$, one gets

$$\begin{aligned}&\frac{1}{V^{\frac{p+q}{2q}}}\frac{\mathrm{d}V}{\mathrm{d}t} = -2\left(\alpha V^{\frac{q-p}{q}} + \beta\right) \\ &\Rightarrow \frac{q}{q-p}\frac{\mathrm{d}V^{\frac{q-p}{2q}}}{\mathrm{d}t} = -\left(\alpha V^{\frac{q-p}{q}} + \beta\right).\end{aligned}$$

Let $z = V^{(q-p)/(2q)}$. One obtains

$$\frac{1}{\alpha z^2 + \beta}\mathrm{d}z = -\frac{q-p}{q}\mathrm{d}t.$$

Integrating both sides of the equation above yields

$$\frac{1}{\sqrt{\alpha\beta}} \arctan\left(\sqrt{\frac{\alpha}{\beta}}z(t)\right) = \frac{1}{\sqrt{\alpha\beta}} \arctan\left(\sqrt{\frac{\alpha}{\beta}}z(0)\right) - \frac{q-p}{q}t.$$

Since the function $\arctan(z) = 0$ if and only if $z = 0$ (implying that $V = 0$), one gets

$$\lim_{t\to T(y_0)} V = 0,$$

where $T(y_0)$ denotes the settling time function given by

$$\begin{aligned}T(y_0) &= \frac{q}{q-p}\frac{1}{\sqrt{\alpha\beta}} \arctan\left(\sqrt{\frac{\alpha}{\beta}}z(0)\right) \\ &= \frac{q}{q-p}\frac{1}{\sqrt{\alpha\beta}} \arctan\left(\sqrt{\frac{\alpha}{\beta}}y_0^{\frac{q-p}{q}}\right).\end{aligned}$$

It can be verified that $T(y_0)$ is bounded by

$$\begin{aligned}\lim_{y_0\to\infty} T(y_0) &= \lim_{y_0\to\infty} \frac{q}{q-p}\frac{1}{\sqrt{\alpha\beta}} \arctan\left(\sqrt{\frac{\alpha}{\beta}}y_0^{\frac{q-p}{q}}\right) \\ &= \frac{q\pi}{2\sqrt{\alpha\beta}(q-p)}.\end{aligned}$$

Note that $V(y(t)) = 0$ implies that $y(t) = 0$. This completes the proof. □

Corollary 2.2 *Consider a scalar system*

$$\dot{x}(t) = -\alpha x^{\frac{m}{n}}(t) - \beta x^{\frac{p}{q}}(t), \quad x(0) = x_0, \tag{2.11}$$

where $\alpha, \beta > 0$, and m, n, p, q satisfying $m > n > 0$ and $q > p > 0$ are odd integers. The equilibrium of system (2.11) is fixed-time stable and the settling time is obtained as

$$T(x_0) < T_{\max} := \frac{1}{\alpha}\frac{n}{m-n} + \frac{1}{\beta}\frac{q}{q-p}. \tag{2.12}$$

Furthermore, if $\varepsilon \triangleq \frac{q(m-n)}{n(q-p)} \leq 1$, a less conservative upper-bound estimate of the settling time is given by

$$T(x_0) < T_{\max} := \frac{q}{q-p}\left(\frac{1}{\sqrt{\alpha\beta}}\arctan\sqrt{\frac{\alpha}{\beta}} + \frac{1}{\alpha\varepsilon}\right). \tag{2.13}$$

Proof Let $V(y) = y^2 \geq 0$. Differentiating $V(y)$ along system (2.11) yields

$$\begin{aligned}\dot{V}(y) &= 2y\left(-\alpha y^{\frac{m}{n}} - \beta y^{\frac{p}{q}}\right)\\ &= -2\alpha(y^2)^{\frac{m+n}{2n}} - 2\beta(y^2)^{\frac{p+q}{2q}}\\ &= -2\left(\alpha V^{\frac{m+n}{2n}-\frac{p+q}{2q}} + \beta\right)V^{\frac{p+q}{2q}}.\end{aligned} \tag{2.14}$$

Since $\alpha V^{(m+n)/2n-(p+q)/(2q)} > 0$, it can be calculated that $\dot{V}(y) \leq -2\beta V^{(p+q)/(2q)}$. In view of $0 < (p+q)/(2q) < 1$, system (2.11) is global finite-time stable by Theorem 2.2. Furthermore, since $V(y) = 0$ in (2.14) is a trivial case, assuming that $V(y) \neq 0$, one obtains

$$\begin{aligned}\frac{1}{V^{\frac{p+q}{2q}}}\frac{\mathrm{d}V}{\mathrm{d}t} &= -2\left(\alpha V^{\frac{m+n}{2n}-\frac{p+q}{q}} + \beta\right)\\ \Rightarrow \frac{q}{q-p}\frac{\mathrm{d}V^{\frac{q-p}{2q}}}{\mathrm{d}t} &= -\left(\alpha V^{\frac{m+n}{2n}-\frac{p+q}{q}} + \beta\right).\end{aligned}$$

Let $z = V^{(q-p)/(2q)}$. One gets

$$\frac{1}{\alpha z^{1+\varepsilon} + \beta}\mathrm{d}z = -\frac{q-p}{q}\mathrm{d}t,$$

where $\varepsilon \triangleq \frac{q(m-n)}{n(q-p)}$. Let $\varphi(z) = \int_0^z \frac{1}{\alpha z^{1+\varepsilon}+\beta}\mathrm{d}z$. Integrating both sides of the equation above yields

$$\varphi(z(t)) = \varphi(z(0)) - \frac{q-p}{q}t.$$

Note that the function $\varphi(z)$ is monotonically increasing. Since $\varphi(z) = 0$ if and only if $z = 0$ (implying that $V = 0$), one obtains

$$\lim_{t \to T(y_0)} V = 0,$$

where $T(y_0)$ denotes the settling time function described by

$$T(y_0) = \frac{q}{q-p}\varphi(z(0)) = \frac{q}{q-p}\varphi(y^{(q-p)/q}(0)).$$

It is clear that $T(y_0)$ is bounded by

$$\begin{aligned}
\lim_{y_0 \to \infty} T(y_0) &= \lim_{z_0 \to \infty} \frac{q}{q-p}\varphi(z(0)) = \frac{q}{q-p}\varphi(\infty) \\
&= \frac{q}{q-p}\left(\int_0^1 \frac{1}{\alpha z^{1+\varepsilon} + \beta}\mathrm{d}z + \int_1^\infty \frac{1}{\alpha z^{1+\varepsilon} + \beta}\mathrm{d}z\right) \\
&< \frac{q}{q-p}\left(\int_0^1 \frac{1}{\beta}\mathrm{d}z + \int_1^\infty \frac{1}{\alpha z^{1+\varepsilon}}\mathrm{d}z\right) \\
&= \frac{q}{q-p}\left(\frac{1}{\beta} + \frac{1}{\alpha\varepsilon}\right) = \frac{1}{\alpha}\frac{n}{m-n} + \frac{1}{\beta}\frac{q}{q-p}.
\end{aligned}$$

Note that $V(y(t)) = 0$ implies that $y(t) = 0$.

In addition, for $0 < \varepsilon \le 1$, a less conservative upper-bound estimate for the settling time, instead of (2.12), can be obtained as

$$\begin{aligned}
\lim_{y_0 \to \infty} T(y_0) &= \frac{q}{q-p}\left(\int_0^1 \frac{1}{\alpha z^{1+\varepsilon} + \beta}\mathrm{d}z + \int_1^\infty \frac{1}{\alpha z^{1+\varepsilon} + \beta}\mathrm{d}z\right) \\
&< \frac{q}{q-p}\left(\int_0^1 \frac{1}{\alpha z^2 + \beta}\mathrm{d}z + \int_1^\infty \frac{1}{\alpha z^{1+\varepsilon}}\mathrm{d}z\right) \\
&= \frac{q}{q-p}\left(\frac{1}{\sqrt{\alpha\beta}}\arctan\sqrt{\frac{\alpha}{\beta}} + \frac{1}{\alpha\varepsilon}\right),
\end{aligned}$$

which completes the proof. □

From Corollaries 2.1 and 2.2, one can see that the settling time function eliminates the dependence of the initial condition x_0. As a result, the convergence time can be prescribed.

As an extension, a Lyapunov criterion [4] for fixed-time stability is obtained in the following theorem. Denote the upper right-hand derivative of a function $\varphi(t)$ by $D^*\varphi(t)$, i.e.,

$$D^*\varphi(t) = \lim_{h \to +0} \sup \frac{\varphi(t+h) - \varphi(t)}{h}.$$

Theorem 2.3 ([4]) *Suppose that there exits a continuous positive definite and radially unbounded function* $V(x) : \mathbb{R}^\ell \to \mathbb{R}^+ \cup \{0\}$ *such that*

$$D^* V(x(t)) \leq -(\alpha V^p(x(t)) + \beta V^q(x(t)))^k \tag{2.15}$$

for α, β, p, q, $k > 0$ *satisfying* $pk < 1$, $qk > 1$. *Then the origin of system (2.5) is fixed-time stable and the settling time is estimated by*

$$T(x_0) \leq T_{\max} := \frac{1}{\alpha^k(1-pk)} + \frac{1}{\beta^k(qk-1)}. \tag{2.16}$$

Proof From (2.15), one obtains

$$\begin{cases} D^* V(x(t)) \leq -\alpha^k V^{pk}(x(t)), & \text{for } V(x(t)) \leq 1, \\ D^* V(x(t)) \leq -\beta^k V^{qk}(x(t)), & \text{for } V(x(t)) > 1. \end{cases}$$

Therefore, for $\forall x_0$ satisfying $V(x_0) > 1$, the last inequality guarantees that $V(x(t)) \leq 1$ for $t \geq \frac{1}{\beta^k(qk-1)}$, and for $\forall x$ such that $V(x(t)) \leq 1$, $V(x(t))$ arrives and stays at the origin for $t \geq t_0 + \frac{1}{\alpha^k(1-pk)}$. Hence, $V(x(t)) = 0$, for $\forall t \geq T_{\max} = \frac{1}{\alpha^k(1-pk)} + \frac{1}{\beta^k(qk-1)}$ and any solution $x(t)$ of (2.5). Since $V(x(t))$ is positive definite and radially unbounded, $x(t) = 0$ for $\forall t \geq T_{\max}$, i.e., the origin is a fixed-time stable equilibrium of system (2.5) with the settling time bounded by (2.16). □

Then, in [7], Theorem 2.3 is refined as the following result.

Theorem 2.4 ([7]) *Suppose that there exits a continuous positive definite and radially unbounded function* $V(x) : \mathbb{R}^\ell \to \mathbb{R}^+ \cup \{0\}$ *such that*

$$D^* V(x(t)) \leq -\alpha V^p(x(t)) - \beta V^q(x(t)) \tag{2.17}$$

for $\alpha, \beta > 0$, $p = 1 - \frac{1}{\mu}$, $q = 1 + \frac{1}{\mu}$, $\mu > 1$. *Then the origin of system (2.5) is fixed-time stable and the settling time estimate is given by*

$$T(x_0) \leq T_{\max} := \frac{\pi\mu}{2\sqrt{\alpha\beta}}. \tag{2.18}$$

Proof The result can be proved by using *Comparison Principle* of differential equations [2]. Consider an auxiliary differential equation given by

$$\dot{y} = -\alpha y^{1-\frac{1}{\mu}} - \beta y^{1+\frac{1}{\mu}}, \quad y_0 = y(0) \geq 0, \tag{2.19}$$

where $\alpha, \beta > 0$, $\mu > 1$. Obviously, $y = 0$ is the equilibrium of (2.19).

Using separation of variables, equation (2.19) can be rewritten as

$$t = -\int_{y_0}^{y} \frac{1}{\alpha y^{1-\frac{1}{\mu}} + \beta y^{1+\frac{1}{\mu}}} \mathrm{d}y.$$

Let $z = y^{1/\mu}$, or equivalently, $y = z^{\mu}$. One gets

$$t = -\mu \int_{z_0}^{z} \frac{z^{\mu-1}}{\alpha z^{\mu-1} + \beta z^{\mu+1}} \mathrm{d}z$$
$$= -\mu \int_{z_0}^{z} \frac{1}{\alpha + \beta z^2} \mathrm{d}z.$$

The solution of (2.19) can be computed as

$$\frac{\mu}{\sqrt{\alpha\beta}} \arctan\left(\sqrt{\frac{\alpha}{\beta}} y^{\frac{1}{\mu}}(t)\right) = -t + \frac{\mu}{\sqrt{\alpha\beta}} \arctan\left(\sqrt{\frac{\alpha}{\beta}} y_0^{\frac{1}{\mu}}\right).$$

Thus, $y(t) = 0$ for all $t \geq T(y_0) := \frac{\mu}{\sqrt{\alpha\beta}} \arctan(\sqrt{\frac{\alpha}{\beta}} y_0^{\frac{1}{\mu}})$. It can be verified that $T(y_0) \leq \frac{\pi\mu}{2\sqrt{\alpha\beta}} = T_{\max}$. This completes the proof. □

2.3 Fixed-Time Stabilization Control

This section addresses a fixed-time stabilization control problem for two classes of systems, which further underscores the feature of fixed-time stability. Two different ways to design fixed-time stabilization controllers are presented, i.e., a terminal sliding mode approach and a homogeneity and Lyapunov approach.

2.3.1 Terminal Sliding Mode Approach

Consider the following nonlinear system in the form of double integrators with a matched perturbation

$$\begin{cases} \dot{x}_1(t) = x_2(t), \\ \dot{x}_2(t) = f(x(t)) + g(x(t))u + d(x(t), t), \end{cases} \quad x(0) = x_0, \tag{2.20}$$

where $x = [x_1, x_2]^T \in \mathbb{R}^2$ is the state vector, $f(x)$ and $g(x) \neq 0$ are smooth vectors on $\mathbb{R}^2$, $u \in \mathbb{R}$ is the control input, and $d(x, t)$ represents the external disturbance satisfying the following assumption.

Assumption 2.3.1 There exists a function $\delta(x, t)$ such that $|d(x, t)| \leq \delta(x, t)$ for all $x \in \mathbb{R}^2$ and $t \geq 0$. □

The control objective is to design a terminal sliding mode control law $u(t)$ such that the origin of system (2.20) is a fixed-time stable equilibrium.

By Corollary 2.2, a terminal sliding surface may be intuitively chosen as

$$s = x_2 + \alpha_1 x_1^{\frac{m_1}{n_1}} + \beta_1 x_1^{\frac{p_1}{q_1}}, \tag{2.21}$$

where m_1, n_1, p_1, and q_1 are positive odd integers satisfying $m_1 > n_1$ and $p_1 < q_1 < 2p_1$, $\alpha_1 > 0$ and $\beta_1 > 0$. With a selection of control input

$$u = -g^{-1}\Big[f + \frac{\alpha_1 m_1}{n_1} x_1^{\frac{m_1}{n_1}-1} x_2 + \frac{\beta_1 p_1}{q_1} x_1^{\frac{p_1}{q_1}-1} x_2 + \alpha_2 s^{\frac{m_2}{n_2}} + \beta_2 s^{\frac{p_2}{q_2}} + \gamma \operatorname{sign}(s)\Big], \tag{2.22}$$

the terminal sliding mode dynamics can be obtained as

$$\dot{s} = -\alpha_2 s^{\frac{m_2}{n_2}} - \beta_2 s^{\frac{p_2}{q_2}} - \gamma(x,t)\operatorname{sign}(s) + d, \tag{2.23}$$

where m_2, n_2, p_2, and q_2 are odd integers satisfying $m_2 > n_2$ and $p_2 < q_2$, $\alpha_2 > 0$, $\beta_2 > 0$ and $\gamma(x,t) = \delta(x,t)$ is the switching gain. Unfortunately, the control calculation in (2.22) is not always well posed since the term containing $x_1^{p_1/q_1-1} x_2$ may cause singularity to occur if $x_1 = 0$ and $x_2 \neq 0$. However, singularity dose not occur on the terminal sliding surface $s = 0$ since $x_1^{p_1/q_1-1} x_2 = -\alpha_1 x_1^{m_1/n_1+p_1/q_1-1} - \beta_1 x_1^{2p_1/q_1-1}$ is nonsingular when $q_1 < 2p_1$. Therefore, the control input (2.22) cannot be guaranteed bounded during the reaching phase.

To circumvent the singularity issue, a variant of the terminal sliding surface in (2.21) is proposed as

$$s = x_1 + \left(\frac{1}{\alpha_1 x_1^{\frac{m_1}{n_1}-\frac{p_1}{q_1}} + \beta_1} x_2\right)^{\frac{q_1}{p_1}} = x_1 + (\kappa x_2)^{\frac{q_1}{p_1}}, \tag{2.24}$$

where $\kappa(x_1) \triangleq 1/(\alpha_1 x_1^{m_1/n_1-p_1/q_1} + \beta_1) > 0$. The terminal sliding surface (2.24) is well defined since $\alpha_1 x_1^{m_1/n_1-p_1/q_1} + \beta_1 > 0$ for all $x_1 \in \mathbb{R}$. It can be straightforwardly verified that (2.24) is equivalent to (2.21) when $s = 0$, implying

$$x_2 = -\alpha_1 x_1^{\frac{m_1}{n_1}} - \beta_1 x_1^{\frac{p_1}{q_1}}. \tag{2.25}$$

Based on (2.24), a fixed-time nonlinear control law is proposed as

$$u = -\frac{[f + \gamma \operatorname{sign}(s)]}{g} + \frac{1}{g\kappa}\left[\alpha_1\left(\frac{m_1}{n_1} - \frac{p_1}{q_1}\right) x_1^{\frac{m_1}{n_1}-\frac{p_1}{q_1}-1} x_2^2 \kappa^2 - \frac{p_1}{q_1}\kappa^{1-\frac{q_1}{p_1}} x_2^{2-\frac{q_1}{p_1}}\right] - \frac{1}{g}\frac{p_1}{q_1}\kappa^{-\frac{q_1}{p_1}} \mu_\tau x_2^{1-\frac{q_1}{p_1}}\left(\alpha_2 s^{\frac{m_2}{n_2}} + \beta_2 s^{\frac{p_2}{q_2}}\right), \tag{2.26}$$

where $\alpha_i > 0, \beta_i > 0, m_i, n_i, p_i, q_i$ are positive odd integers satisfying $m_i > n_i, (i = 1, 2)$, $p_1 < q_1 < 2p_1$, $p_2 < q_2$ and $m_1/n_1 - p_1/q_1 > 1$ and $\mu_\tau(\cdot) : [0, +\infty) \to [0, 1]$ is a C^1 function, given by

$$\mu_\tau(x_2^{q_1/p_1-1}) = \begin{cases} \sin\left(\frac{\pi}{2} \cdot \frac{x_2^{q_1/p_1-1}}{\tau}\right) & \text{if } x_2^{\frac{q_1}{p_1}-1} \leq \tau; \\ 1 & \text{otherwise,} \end{cases} \tag{2.27}$$

for some constant τ. The fact that $\sin \omega x \leq x$, $(\omega > 0)$ ensures that the control law (2.26) is always well defined with proper selection of the corresponding control parameters (i.e., $q_1/p_1 < 2$ and $m_1/n_1 - p_1/q_1 > 1$).

Theorem 2.5 *Consider the second-order system (2.20) with the feedback control law designed as (2.26). Then the states $x = [x_1, x_2]^T$ converge to the origin within fixed time and the settling time estimate is derived as*

$$T < T_{\max} \triangleq T_1 + T_2 + \epsilon(\tau), \tag{2.28}$$

where

$$T_1 = \frac{1}{\alpha_1}\frac{n_1}{m_1 - n_1} + \frac{1}{\beta_1}\frac{q_1}{q_1 - p_1}, \quad T_2 = \frac{1}{\alpha_2}\frac{n_2}{m_2 - n_2} + \frac{1}{\beta_2}\frac{q_2}{q_2 - p_2},$$

and $\epsilon(\tau)$ is a small time margin related to the boundary width $2\tau^{p_1/(q_1-p_1)}$.

Proof Differentiating s in (2.24) against time yields

$$\dot{s} = x_2 + \frac{q_1}{p_1}(\kappa x_2)^{\frac{q_1}{p_1}-1}\bigg[-\alpha_1\left(\frac{m_1}{n_1} - \frac{p_1}{q_1}\right)x_1^{\frac{m_1}{n_1}-\frac{p_1}{q_1}-1}x_2^2\kappa^2 \\ + \kappa(f + gu + d)\bigg]. \tag{2.29}$$

Substituting (2.26) into (2.29), one obtains

$$\dot{s} = -\mu_\tau\left(\alpha_2 s^{\frac{m_2}{n_2}} + \beta_2 s^{\frac{p_2}{q_2}}\right) - \frac{q_1}{p_1}\kappa^{\frac{q_1}{p_1}}x_2^{\frac{q_1}{p_1}-1}\left[d + \gamma\text{sign}(s)\right]. \tag{2.30}$$

Note that $\mu_\tau > 0$ if $x_2 \neq 0$. For the convenience of the proof, the state space $x \in \mathbb{R}^2$ is divided into two different areas

$$\mathscr{S}_1 = \{(x_1, x_2) | x_2^{q_1/p_1-1} \geq \tau\} \quad \text{and} \quad \mathscr{S}_2 = \{(x_1, x_2) | x_2^{q_1/p_1-1} < \tau\},$$

as shown in Fig. 2.1. Consider the Lyapunov function $V_2 = s^2/2$ and its time derivative can be derived as

$$\begin{aligned}
\dot{V}_2 &\le -\mu_\tau \alpha_2 s^{\frac{m_2+n_2}{n_2}} - \mu_\tau \beta_2 s^{\frac{p_2+q_2}{q_2}} + \frac{q_1}{p_1}\kappa^{\frac{q_1}{p_1}} x_2^{\frac{q_1}{p_1}-1}(|d|-\gamma)|s| \\
&\le -\mu_\tau \alpha_2 (s^2)^{\frac{m_2+n_2}{2n_2}} - \mu_\tau \beta_2 (s^2)^{\frac{p_2+q_2}{2q_2}} \\
&= -\mu_\tau \alpha_2 (2V_2)^{\frac{m_2+n_2}{2n_2}} - \mu_\tau \beta_2 (2V_2)^{\frac{p_2+q_2}{2q_2}}.
\end{aligned} \tag{2.31}$$

(i) When the system states (x_1, x_2) are in the area $\mathscr{S}_1$, the function $\mu_\tau(\cdot)$ takes value one, i.e., $\mu_\tau = 1$. Then $\dot{s} = -\alpha_2 s^{m_2/n_2} - \beta_2 s^{p_2/q_2} - \frac{q_1}{p_1}\kappa^{q_1/p_1} x_2^{q_1/p_1-1}[d + \gamma \text{sign}(s)]$ follows and thus

$$\dot{V}_2 \le -\alpha_2 (2V_2)^{(m_2+n_2)/2n_2} - \beta_2 (2V_2)^{(p_2+q_2)/2q_2}.$$

If $V_2 \ne 0$, then let $y = \sqrt{2V_2}$ be the solution to the differential equation (2.11), where $\dot{y} = \dot{V}_2/\sqrt{2V_2}$ is used. Note that $V_2 = 0$ implies that $s = 0$. It follows from Corollary 2.2 and *Comparison Principle* of differential equations [2] that the states (x_1, x_2) will reach the sliding surface $s = 0$ or enter the area $\mathscr{S}_2$ within fixed-time $t_1 < T_2$.

(ii) In the area $\mathscr{S}_2$, $0 < \mu_\tau < 1$ when $x_2 \ne 0$. It can be verified similarly from (2.31) that the terminal sliding surface $s = 0$ is still an attractor. What remains is to prove that the x_1-axis in Fig. 2.1 is not attractive except for the origin $(x_1, x_2) = (0, 0)$. It can be shown that on the x_1-axis the control input (2.26) degenerates into the following form:

$$u = -\frac{[f + \gamma \text{sign}(s)]}{g} - \frac{1}{g}\frac{p_1}{q_1}\kappa^{-\frac{q_1}{p_1}}\left(\alpha_2 s^{\frac{m_2}{n_2}} + \beta_2 s^{\frac{p_2}{q_2}}\right), \tag{2.32}$$

where the fact $\mu_\tau x_2^{1-p_1/q_1} \to \pi/2\tau$ when $x_2 \to 0$ is used. Substituting (2.32) into the second equation of (2.20) yields

$$\dot{x}_2 = d - \gamma \text{sign}(s) - \frac{p_1}{q_1}\kappa^{-\frac{q_1}{p_1}}\left(\alpha_2 s^{\frac{m_2}{n_2}} + \beta_2 s^{\frac{p_2}{q_2}}\right). \tag{2.33}$$

In view of $\gamma \ge |d|$ and $\kappa > 0$, one gets that $\dot{x}_2 < 0$ for $s > 0$ and $\dot{x}_2 > 0$ for $s < 0$. Thus, the trajectory $x(t)$ will transgress the area $\mathscr{S}_2$ into $\mathscr{S}_1$ monotonically in finite time, as shown in Fig. 2.1. Then, for a given $\tau > 0$, there exists a constant $\epsilon(\tau) > 0$ specifying the time margin for the system trajectory to pass through area $\mathscr{S}_2$ where the sinusoidal function is activated.

It can be concluded that the sliding surface $s = 0$ can be reached from anywhere in the phase plane within fixed time $t_1 < T_2 + \epsilon(\tau)$, where $\epsilon(\tau)$ is determined by the boundary width $2\tau^{p_1/(q_1-p_1)}$. Once the sliding surface $s = 0$ is reached, it follows from (2.25) and Corollary 2.2 that the states (x_1, x_2) will reach the origin within fixed-time $t_2 < T_1$. Hence, the settling time T is bounded by (2.28). □

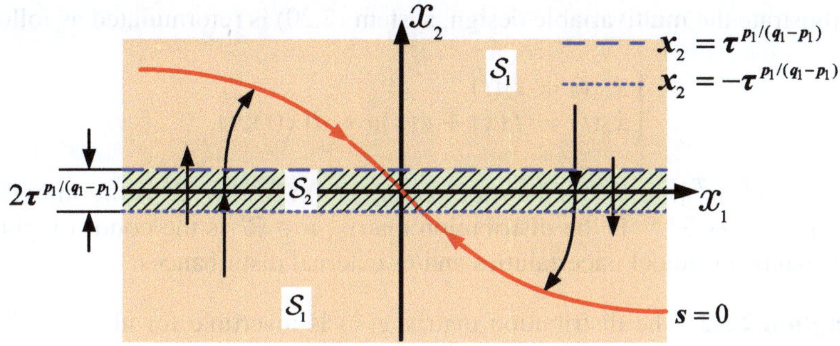

Fig. 2.1 The phase portrait of the system

Remark 2.1 The TSMC law (2.26) is undefined in the numerical implementation when the initial velocity $x_2(0) = 0$. To address this concern, a small perturbation to the initial velocity is required in practice. Alternatively, a modified control input is proposed as follows:

$$u = \begin{cases} -\frac{[f+\gamma \mathrm{sign}(s)]}{g} + \frac{1}{g\kappa}\left[\alpha_1\left(\frac{m_1}{n_1}-\frac{p_1}{q_1}\right)x_1^{\frac{m_1}{n_1}-\frac{p_1}{q_1}-1}x_2^2\kappa^2 - \frac{p_1}{q_1}\kappa^{1-\frac{q_1}{p_1}}x_2^{2-\frac{q_1}{p_1}}\right] \\ \quad -\frac{1}{g}\frac{p_1}{q_1}\kappa^{-\frac{q_1}{p_1}}\left(\alpha_2 s^{\frac{m_2}{n_2}}+\beta_2 s^{\frac{p_2}{q_2}}\right), \qquad x_2(0)=0; \\ -\frac{[f+\gamma \mathrm{sign}(s)]}{g} + \frac{1}{g\kappa}\left[\alpha_1\left(\frac{m_1}{n_1}-\frac{p_1}{q_1}\right)x_1^{\frac{m_1}{n_1}-\frac{p_1}{q_1}-1}x_2^2\kappa^2 - \frac{p_1}{q_1}\kappa^{1-\frac{q_1}{p_1}}x_2^{2-\frac{q_1}{p_1}}\right] \\ \quad -\frac{1}{g}\frac{p_1}{q_1}\kappa^{-\frac{q_1}{p_1}}\mu_\tau x_2^{1-\frac{q_1}{p_1}}\left(\alpha_2 s^{\frac{m_2}{n_2}}+\beta_2 s^{\frac{p_2}{q_2}}\right), \quad x_2(0)\neq 0. \end{cases}$$

Remark 2.2 It is worthwhile to note that the traveling time $\epsilon(\tau)$ across the area $\mathscr{S}_2$ cannot be estimated from theoretical viewpoint due to the introduction of a boundary layer to activate the sinusoidal function and avoid the occurrence of singularity. However, the conservative estimate of T_2 justifies neglecting $\epsilon(\tau)$ in practice for small τ.

Remark 2.3 Likewise, if $\varepsilon_1 \triangleq \frac{q_1(m_1-n_1)}{n_1(q_1-p_1)} \leq 1$ and/or $\varepsilon_2 \triangleq \frac{q_2(m_2-n_2)}{n_2(q_2-p_2)} \leq 1$, a less conservative bound (2.13) can be used for calculating T_1 or T_2. However, it can be verified that the parameter constraints $m_1/n_1 - p_1/q_1 > 1$ and $p_1/q_1 > 1/2$ lead to $\varepsilon_1 > 1$. Thus, the estimate (2.13) is only suitable for computing tighter T_2 since the case $\varepsilon_2 \leq 1$ may be possible.

For multivariable second-order systems, an extension of Theorem 2.5 is also applicable. In other words, the decoupled design is usually exploited to formulate a multivariable control problem involving single-input control channels. However, different from the decoupled design, we will present a nonlinear coupled sliding mode control method. With the coupled injection terms, the input discontinuity and the singularity issue are significantly reduced, which only occur when all elements of the related vector are zero.

To illustrate the multivariable design, system (2.20) is reformulated as follows:

$$\begin{cases} \dot{x}_1(t) = x_2(t), \\ \dot{x}_2(t) = f(x) + g(x)u + d(x(t), t), \end{cases} \tag{2.34}$$

where $x = [x_1^T, x_2^T]^T \in \mathbb{R}^{2\ell}$ is the state vector, $f(x) : \mathbb{R}^{2\ell} \to \mathbb{R}^{\ell}$ is the smooth vector, $g(x) : \mathbb{R}^{2\ell} \to \mathbb{R}^{\ell\times\ell}$ is the distribution matrix, $u \in \mathbb{R}^{\ell}$ is the control input, and $d(x, t)$ stands for model uncertainties and/or external disturbances.

Assumption 2.3.2 The distribution matrix $g(x)$ is invertible for all $x \in \mathbb{R}^{2\ell}$, i.e., $g^{-1}(x)$ exists. □

Assumption 2.3.3 There exists a known function $\delta(x, t) : \mathbb{R}^{2\ell} \times \mathbb{R}_+ \to \mathbb{R}$ such that $\|d(x, t)\| \le \delta(x, t)$, $\forall x \in \mathbb{R}^{2\ell}$ and $\forall t \ge 0$. □

Before proceeding to control design for system (2.34), we first consider a fixed-time first-order multivariable system as a foundation for the second-order case.

Lemma 2.1 *Consider a first-order multivariable system*

$$\dot{x} = -\alpha x\|x\|^{p-1} - \beta x\|x\|^{q-1}, \tag{2.35}$$

where $x \in \mathbb{R}^{\ell}$, $\alpha, \beta > 0$, $p > 1$ *and* $0 < q < 1$. *Then the system is globally fixed-time stable at the origin and the settling time estimate is given by* (2.16) *with* $k = 1$.

Proof System (2.35) can be rewritten as

$$\dot{x} = -\alpha\frac{x}{\|x\|}\|x\|^{p} - \beta\frac{x}{\|x\|}\|x\|^{q}, \tag{2.36}$$

from which one can see that system (2.36) is well defined since $\|x\| = 0 \Leftrightarrow x = 0$.

Consider the following Lyapunov function:

$$V(x) = \frac{1}{2}x^{\mathrm{T}}x.$$

Differentiating V along the trajectory of system (2.36) yields

$$\begin{aligned} D^*V(x) &= -\alpha x^T x\|x\|^{p-1} - \beta x^T x\|x\|^{q-1} \\ &= -\alpha\|x\|^{p+1} - \beta\|x\|^{q+1} \\ &= -\alpha(2V)^{\frac{p+1}{2}} - \beta(2V)^{\frac{q+1}{2}}. \end{aligned} \tag{2.37}$$

Let $y = \sqrt{2V} \ge 0$. We have that $\dot{y} = \dot{V}/\sqrt{2V}$. It follows from (2.37) that $\dot{y} = -\alpha y^p - \beta y^q$. Thus, by Theorem 2.3 with $k = 1$, system (2.36) is globally fixed-time stable at the origin and the settling time is bounded by

$$T(x_0) \le T_{\max} := \frac{1}{\alpha(p-1)} + \frac{1}{\beta(1-q)},$$

which completes the proof.

□

Remark 2.4 Compared with single variable SMC in some existing results, discontinuity only occurs when all the components of $x = [x_1, x_2, \ldots, x_\ell]^T$ equal to zero simultaneously, i.e., $x_i = 0$ for all $i = 1, 2, \ldots, \ell$. Therefore, the proposed structure has a better property of chattering reduction.

For fixed-time stabilization of system (2.34), the following nonlinear coupled sliding surface is proposed:

$$s = x_2 + \alpha_2 x_1 \|x_1\|^{p_2-1} + \beta_2 x_1 \|x_1\|^{q_2-1}, \tag{2.38}$$

where $\alpha_2, \beta_2 > 0$, $p_2 > 1$, and $1/2 < q_2 < 1$.

Then we take the control input in the following form:

$$\begin{aligned} u = & -\alpha_2 g^{-1}(x)\left[I + (p_2-1)\frac{x_1 x_1^T}{\|x_1\|}\right]\|x_1\|^{p_2-1}x_2 \\ & -\beta_2 g^{-1}(x)\left[I + (q_2-1)\frac{x_1 x_1^T}{\|x_1\|}\right]\|x_1\|^{q_2-1}x_2 \\ & -g^{-1}(x)\left[f(x) + \rho(x,t)\frac{s}{\|s\|}\right] - g^{-1}(x)\left[\alpha_1 s\|s\|^{p_1-1} + \beta_1 s\|s\|^{q_1-1}\right], \end{aligned} \tag{2.39}$$

where $\alpha_1, \beta_1 > 0$, $p_1 > 1, 0 < q_1 < 1$, and $\rho(x,t) \ge \delta(x,t)$.

Theorem 2.6 *Consider the second-order multivariable nonlinear system* (2.34) *subject to the state feedback control input* (2.39). *If Assumptions 2.3.2 and 2.3.3 hold, then the closed-loop system is globally fixed-time stable at the origin and the settling time estimate is given by*

$$T(x_0) \le T_{\max} := \sum_{i=1}^{2}\left[\frac{1}{\alpha_i(p_i-1)} + \frac{1}{\beta_i(1-q_i)}\right]. \tag{2.40}$$

Proof (i) With (2.34) and (2.38), the time derivative of s can be computed as

$$\begin{aligned} \dot{s} = & f(x) + \delta_f(x) + g(x)u + \delta_d(x,t) \\ & + \alpha_2\left[I + (p_2-1)\frac{x_1 x_1^{\mathrm{T}}}{\|x_1\|}\right]\|x_1\|^{p_2-1}x_2 \\ & + \beta_2\left[I + (q_2-1)\frac{x_1 x_1^{\mathrm{T}}}{\|x_1\|}\right]\|x_1\|^{q_2-1}x_2. \end{aligned} \tag{2.41}$$

Substituting (2.39) into (2.41) yields

$$\begin{aligned}\dot{s} = & -\alpha_1 s\|s\|^{p_1-1} - \beta_1 s\|s\|^{q_1-1} + d(x,t)\\ & - \rho(x,t)\frac{s}{\|s\|}.\end{aligned} \tag{2.42}$$

Choosing $V(s) = s^{\mathrm{T}}s/2$ and differentiating, it can be verified that

$$\begin{aligned}D^*V(s) \leq & -\alpha_1\|s\|^{p_1+1} - \beta_1\|s\|^{q_1+1}\\ & - [\rho(x,t) - \|d(x,t)\|]\,\|s\|\\ \leq & -\alpha_1(2V)^{\frac{p_1+1}{2}} - \beta_1(2V)^{\frac{q_1+1}{2}}.\end{aligned} \tag{2.43}$$

Let $y = \sqrt{2V}$. A similar technique can be employed as in Lemma 2.1, and we have that $\dot{y} \leq -\alpha y^p - \beta y^q$ which is globally fixed-time stable at the origin by Theorem 2.3. By using Comparison Principle, s approaches zero with a finite settling time bounded by

$$T_1 := \frac{1}{\alpha_1(p_1-1)} + \frac{1}{\beta_1(1-q_1)}. \tag{2.44}$$

(ii) For all $t \geq T_1$, the ideal sliding motion is achieved, i.e., $s = 0$. It follows from (2.38) that the corresponding reduced-order motion is governed by

$$\dot{x}_1 = -\alpha_2 x_1\|x_1\|^{p_2-1} - \beta_2 x_1\|x_1\|^{q_2-1}. \tag{2.45}$$

From Lemma 2.1, fixed-time stability of system (2.45) can be guaranteed and the settling time estimate is given by

$$T_2 := \frac{1}{\alpha_2(p_2-1)} + \frac{1}{\beta_2(1-q_2)}. \tag{2.46}$$

Thus, the closed-loop system (2.34) with the input (2.39) is globally fixed-time stable at the origin with the settling time function bounded by $T(x_0) \leq T_1 + T_2$. This completes the proof. □

Note that the control input is not well defined in the set $\mathscr{S} = \{(x_1, x_2)|x_1 = 0, x_2 \neq 0\}$, where input singularity occurs due to the term $\|x_1\|^{q_2-1}x_2 = \|x_1\|^{q_2}x_2/\|x_1\|$ in (2.39) with $q_2 < 1$. Further observation reveals that the singularity occurs only if all the components of $x_1 = [x_{11}, x_{12}, \ldots, x_{1\ell}]^T$ are identically zero, i.e., $x_{1j} = 0$ for all $j = 1, 2, \ldots, \ell$. Therefore, the multivariable fixed-time stabilization control has a better property of singularity reduction over the single variable case. However, the control input is well defined during the sliding motion if $1/2 < q_2 < 1$, since the control input degenerates into

$$
\begin{aligned}
u = & -\alpha_2 g^{-1}(x)\left[I + (p_2 - 1)\frac{x_1 x_1^{\mathrm{T}}}{\|x_1\|}\right]\left(\alpha_2\|x_1\|^{2p_2-1} + \alpha_2\|x_1\|^{p_2+q_2-1}\right)\frac{x_1}{\|x_1\|} \\
& - \beta_2 g^{-1}(x)\left[I + (q_2 - 1)\frac{x_1 x_1^{\mathrm{T}}}{\|x_1\|}\right]\left(\beta_2\|x_1\|^{2q_2-1} + \alpha_2\|x_1\|^{p_2+q_2-1}\right)\frac{x_1}{\|x_1\|} \\
& - g^{-1}(x)\left[f(x) + \rho(x,t)\frac{s}{\|s\|}\right] - g^{-1}(x)\left[\alpha_1 s\|s\|^{p_1-1} + \beta_1 s\|s\|^{q_1-1}\right],
\end{aligned}
$$

where the fact that $s = 0 \Leftrightarrow x_2 = -\alpha_2 x_1\|x_1\|^{p_2-1} - \beta_2 x_1\|x_1\|^{q_2-1}$ has been used. Thus, one can avoid the occurrence of the singularity by appropriately prescribing the system initial conditions.

To avoid the singularity for implementation of the proposed controller, an alternative way is to introduce the following saturation into (2.39)

$$
\operatorname{sat}_\epsilon(x) = \operatorname{sign}(x)\cdot\min\{|x|, \epsilon\}, \tag{2.47}
$$

where $x \in \mathbb{R}$ and ϵ is a threshold parameter. With (2.47), a practical control input can be formulated as

$$
\begin{aligned}
u = & -\alpha_2 g^{-1}(x)\left[I + (p_2 - 1)\frac{x_1 x_1^{\mathrm{T}}}{\|x_1\|}\right]\|x_1\|^{p_2-1}x_2 \\
& - \beta_2 g^{-1}(x)\left[I + (q_2 - 1)\frac{x_1 x_1^{\mathrm{T}}}{\|x_1\|}\right]\|x_1\|^{q_2}\operatorname{sat}_\epsilon\left(\frac{x_2}{\|x_1\|}\right) \\
& - g^{-1}(x)\left[f(x) + \rho(x,t)\frac{s}{\|s\|}\right] \\
& - g^{-1}(x)\left[\alpha_1 s\|s\|^{p_1-1} + \beta_1 s\|s\|^{q_1-1}\right],
\end{aligned} \tag{2.48}
$$

where the saturation function $\operatorname{sat}(\cdot)$ is applied componentwise, i.e.,

$$
\operatorname{sat}_\epsilon\left(\frac{x_2}{\|x_1\|}\right) = \left[\operatorname{sat}_\epsilon\left(\frac{x_{21}}{\|x_1\|}\right), \ldots, \operatorname{sat}_\epsilon\left(\frac{x_{2\ell}}{\|x_1\|}\right)\right]^{\mathrm{T}}
$$

with $x_2 = [x_{21}, x_{22}, \ldots, x_{2\ell}]^{\mathrm{T}}$.

Theorem 2.7 *Consider the second-order multivariable nonlinear system* (2.34) *subject to the state feedback control input* (2.48). *If Assumptions 2.3.2 and 2.3.3 hold, then the closed-loop system is globally almost fixed-time stable at the origin and the settling time estimate is given by*

$$
T(x_0) \le T_{\max} := \sum_{i=1}^{2}\left[\frac{1}{\alpha_i(p_i - 1)} + \frac{1}{\beta_i(1 - q_i)}\right] + T_\epsilon, \tag{2.49}
$$

where T_ϵ is a small time margin to account for the influence of the saturation manipulation.

Proof With the same sliding surface defined in (2.38) and the corresponding time derivative (2.41), substituting the control input (2.48) into (2.41) yields

$$\begin{aligned}\dot{s} =& -\alpha_1 s\|s\|^{p_1-1} - \beta_1 s\|s\|^{q_1-1} + d(x,t) - \rho(x,t)\frac{s}{\|s\|} \\ &+ \beta_2\left[I + (q_2-1)\frac{x_1x_1^T}{\|x_1\|}\right]\|x_1\|^{q_2}\left[\frac{x_2}{\|x_1\|} - \text{sat}_\epsilon\left(\frac{x_2}{\|x_1\|}\right)\right]. \end{aligned} \tag{2.50}$$

To show the sliding surface is globally attractive, several subsets of $(x_1, x_2) \in \mathbb{R}^{2n}$, corresponding to the occurrence of saturation, are defined for the convenience of discussion:

$$\mathcal{S}_i = \left\{(x_1, x_{2i}) \left| \frac{|x_{2i}|}{\|x_1\|} \geq \epsilon \right.\right\}, \quad i = 1, 2, \ldots, \ell. \tag{2.51}$$

Let

$$\mathcal{R} = \mathcal{S} - \mathcal{T}, \quad \mathcal{T} = \mathcal{S}_1 \cup \mathcal{S}_1 \cup \cdots \cup \mathcal{S}_\ell.$$

(i) When the system states are in the area $\mathcal{R}$, we have that $\text{sat}_\epsilon(x_2/\|x_1\|) = x_2/\|x_1\|$ and (2.50) can be written as (2.42). Applying the same technique as the one in Theorem 2.6, we have that $s(t) = 0$ for all $t > T_1$ with T_1 defined in (2.44), or the system trajectory enters the area $\mathcal{T}$ in finite time bounded by T_1.

(ii) When the system states are in the area $\mathcal{T}$, we have that

$$\text{sat}_\epsilon\left(\frac{x_{2k}}{\|x_1\|}\right) = \epsilon, \quad \exists k \in \{1, 2, \ldots, \ell\},$$

which implies that one or some elements of the control input vector $u = [u_1, u_2, \ldots, u_\ell]^T$ is or are saturated due to the small value of $\|x_1\|$. As shown in Fig. 2.2, there are only two cases to be considered for each sliding surface: $x_{2k} > 0$ and $x_{2k} < 0$.

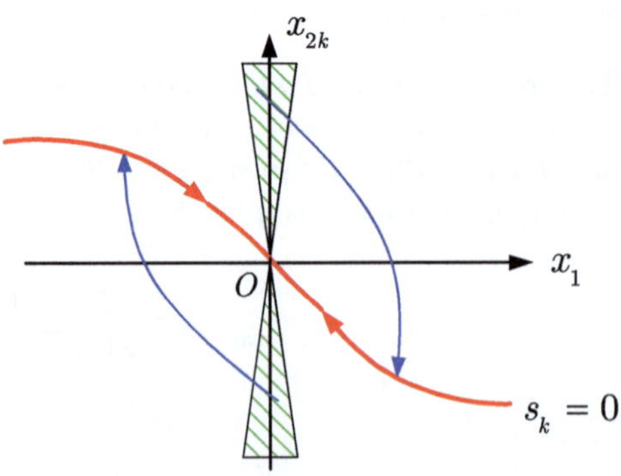

Fig. 2.2 Phase portrait: the k-th sliding surface

From (2.34), the solution for x_{1k} can be computed as

$$x_{1k}(t) = x_{1k}(0) + \int_0^t x_{2k}(\tau)\mathrm{d}\tau.$$

It is clear to see that x_{1k} will monotonically increase or decrease until hitting the boundary of $\mathscr{S}_k$ in a single direction. Thus, for any given ϵ, there exists a finite-time T_{ϵ_k} such that the system trajectory transgresses $\mathscr{S}_k$ and $|x_{2k}|/\|x_1\| < \epsilon$, i.e., $\mathrm{sat}_\epsilon(x_{2k}/\|x_1\|) = x_{2k}/\|x_1\|$. The same holds for all $\mathrm{sat}_\epsilon(x_{2j}/\|x_1\|) = x_{2j}/\|x_1\|$ with $j \neq k$. At the worst case, for a given ϵ, there exists a finite-time $T_\epsilon \leq \sum_{i=1}^n T_{\epsilon_i}$ such that the system trajectories leave the area $\mathscr{T}$.

(iii) For all $t \geq T_1 + T_\epsilon$, the ideal sliding motion (2.45) is induced, i.e., $s = 0$. Similar to the part (ii) in the proof of Theorem 2.6, fixed-time stability of (2.45) can be guaranteed and the settling time estimate T_2 is given in (2.46). Thus, the closed-loop system (2.34) with the input (2.48) is globally *almost* fixed-time stable at the origin with the settling time function bounded by $T(x_0) \leq T_1 + T_2 + T_\epsilon$. This completes the proof. □

Remark 2.5 Applying Theorem 2.4, if $p_i = 1 + \frac{1}{\mu_i}$ and $q_i = 1 - \frac{1}{\mu_i}$ with $\mu_i > 1$, $i = 1, 2$, less conservative estimates for the settling time upper bound in Theorems 2.6 and 2.7 can be computed, respectively, as

$$T(x_0) \leq T_{\max} := \sum_{i=1}^{2} \frac{\pi\mu_i}{2\sqrt{\alpha_i\beta_i}}, \tag{2.52}$$

$$T(x_0) \leq T_{\max} := \sum_{i=1}^{2} \frac{\pi\mu_i}{2\sqrt{\alpha_i\beta_i}} + T_\epsilon. \tag{2.53}$$

Remark 2.6 If a sufficiently large threshold parameter ϵ is chosen, then $\|x_1\|$ is allowed to be very small and thus the area $\mathscr{T}$ tends to be very small. Therefore, the time margin T_ϵ can be made very small. In general, the settling time estimates T_1 and T_2 for global fixed-time stability are very conservative. In practice, T_ϵ can be negligible. Thus, we use the word "almost" to clarify this issue in Theorem 2.7.

It should be pointed out that an extension of Theorems 2.5–2.7 to higher order nonlinear systems ($\ell > 2$) is prohibitive due to a singularity issue. The difficulties arise from the recursive control structure that requires successive differentiation of the sliding variable in each level which contains the term with fractional exponent. Thus, the technique used in Theorem 2.5 is incapable of dealing with more than twice differentiation of the sliding variable proposed in the first level. Note, however, that a homogeneity technique provides an alternative way to overcome this difficulty, which is a typical non-recursive design approach.

2.3.2 Homogeneity and Lyapunov Approach

The autonomous system (2.5) with relative degree ℓ can be converted into an ℓ-th-order integrator system by a proper selection of the output and a successive differentiation.

Consider an ℓth-order ($\ell \geq 2$) integrator system

$$\dot{x}(t) = Ax(t) + Bu(t), \quad x(0) = x_0, \tag{2.54}$$

$$A = \begin{bmatrix} 0 & 1 & \cdots & 0 \\ \vdots & \vdots & \ddots & \vdots \\ 0 & 0 & \cdots & 1 \\ 0 & 0 & \cdots & 0 \end{bmatrix} \quad \text{and} \quad B = \begin{bmatrix} 0 \\ \vdots \\ 0 \\ 1 \end{bmatrix},$$

where $x(t) = [x_1, x_2, \ldots, x_\ell]^T \in \mathbb{R}^\ell$ denotes the state and $u(t) \in \mathbb{R}$ stands for the control input.

Let $r = [r_1, r_2, \ldots, r_\ell]^T \in \mathbb{R}^\ell$ be the weight vector with $r_i > 0$. The dilation mapping is defined by $\Lambda_\lambda^r(x) = [\lambda^{r_1} x_1, \lambda^{r_2} x_2, \ldots, \lambda^{r_\ell} x_\ell]^T$ for $\forall \lambda > 0$ and $\forall x \in \mathbb{R}^\ell$.

Definition 2.5 ([8]) A function $f_i(x)$ is said to be r-homogeneous with degree $\kappa \in \mathbb{R}$ if for $\forall x \in \mathbb{R}^\ell$ and $\forall \lambda > 0$, $f_i(\Lambda_\lambda^r(x)) = \lambda^\kappa f_i(x)$ holds.

Definition 2.6 ([8]) A vector field $f(x) = [f_1(x), f_2(x), \ldots, f_\ell(x)]^T$ is said to be r-homogeneous with degree $\kappa \in \mathbb{R}$ if the component $f_i(x)$ is r-homogeneous with degree of $\kappa + r_i$, i.e., $f_i(\Lambda_\lambda^r(x)) = \lambda^{\kappa + r_i} f_i(x)$ for $\forall \lambda > 0$.

The following two useful lemmas are recalled for stability analysis.

Lemma 2.2 *([8]) Suppose that system* (2.5) *is homogeneous with negative degree* $\kappa < 0$ *with respect to dilation* $\Lambda_\lambda^r(x)$. *If the origin is an asymptomatically stable equilibrium of the system* (2.5), *then the origin is globally finite-time stable.*

Lemma 2.3 *([8]) Suppose that* V_1 *and* V_2 *are continuous real-valued functions on* $\mathbb{R}^\ell$, *homogeneous of degrees* $\kappa_1 > 0$ *and* $\kappa_2 > 0$, *respectively, and* V_1 *is positive definite. Then, for every* $x \in \mathbb{R}^\ell$,

$$\Big[\min_{\{z:V_1(z)=1\}} V_2(z)\Big] V_1^{\frac{\kappa_2}{\kappa_1}}(x) \leq V_2(x) \leq \Big[\max_{\{z:V_1(z)=1\}} V_2(z)\Big] V_1^{\frac{\kappa_2}{\kappa_1}}(x). \tag{2.55}$$

A fixed-time homogeneous stabilization control law takes the following form:

$$u(t) = v(t) + w(t), \tag{2.56}$$

where

$$v(t) = -\sum_{i=1}^{\ell} k_i |x_i|^{\gamma_i} \operatorname{sign}(x_i), \tag{2.57}$$

$$w(t) = -\sum_{i=1}^{\ell} \bar{k}_i |x_i|^{\bar{\gamma}_i} \operatorname{sign}(x_i) \tag{2.58}$$

with parameters γ_i and $\bar{\gamma}_i$ satisfying, for $i = 2, 3, \ldots, \ell$,

$$\gamma_{i-1} = \frac{\gamma_i \gamma_{i+1}}{2\gamma_{i+1} - \gamma_i} \quad \text{and} \quad \bar{\gamma}_{i-1} = \frac{\bar{\gamma}_i \bar{\gamma}_{i+1}}{2\bar{\gamma}_{i+1} - \bar{\gamma}_i}, \tag{2.59}$$

where $\gamma_{\ell+1} = \bar{\gamma}_{\ell+1} = 1$, $\gamma_\ell = \gamma_0 \in (1-\epsilon, 1)$ and $\bar{\gamma}_\ell = \bar{\gamma}_0 \in (1, 1+\epsilon)$ for a sufficiently small $\epsilon > 0$.

Obviously, the control structure (2.56) is a standard fixed-time stabilization structure that generally consists of two classes of feedback terms: one with fractional exponent and the other with exponent greater than 1. For high-order systems ($\ell > 2$), the construction of Lyapunov functions for stability analysis is not straightforward. The construction of Lyapunov functions is provided in [9] for fixed-time stability analysis and a corresponding estimate for settling time is derived.

Theorem 2.8 ([9]) *Consider a dynamic system* (2.54) *with* (2.56). *The state* $x(t) = [x_1(t), x_2(t), \ldots, x_\ell(t)]^T \in \mathbb{R}^\ell$ *converges to the origin for a fixed time*

$$T \le T_{\max} := \frac{\lambda_{\max}^{\rho}(P_1)}{r_1 \rho} + \frac{1}{r_2 \sigma \Upsilon^{\sigma}},$$

where $\rho = (1-\gamma_0)/\gamma_0$, $\sigma = (\bar{\gamma}_0 - 1)/\bar{\gamma}_0$, $r_i = \lambda_{\min}(Q_i)/\lambda_{\max}(P_i)$, $i = 1, 2$, $\Upsilon \le \lambda_{\min}(P_2)$ *is a positive number,* P_i *is a solution to a Lyapunov function* $P_i A_i + A_i^T P_i = -Q_i$ *with* $Q_i \in \mathbb{R}^{\ell\times\ell}$ *being a symmetric positive definite matrix and* A_i *being a Hurwitz matrix defined in the sequel.*

For more clarity and easier understanding, we will rephrase the proof given in [9] and prove the stability in several steps.

First, consider that the system (2.54) subject to the control input (2.57) and the resultant closed-loop system takes the form of

$$\dot{x}(t) = Ax(t) - B\Big(\sum_{i=1}^{\ell} k_i |x_i|^{\gamma_i} \operatorname{sign}(x_i)\Big). \tag{2.60}$$

Define the Lyapunov function candidate for system (2.60) as $V_1(\zeta) = \zeta^T P_1 \zeta$ with $\zeta = [x_1^{\gamma_1}, x_2^{\gamma_2}, \ldots, x_\ell^{\gamma_\ell}]^T$ and $P_1 = P_1^T > 0$ is the solution of algebraic Lyapunov function $P_1 A_1 + A_1^T P_1 = -Q_1$ for arbitrary $Q_1 > 0$ and A_1 is in the controllable canonical form:

$$A_1 = \begin{bmatrix} 0 & 1 & \cdots & 0 \\ \vdots & \vdots & \ddots & \vdots \\ 0 & 0 & \cdots & 1 \\ -k_1 & -k_2 & \cdots & -k_\ell \end{bmatrix},$$

where $k_i > 0$, $(i = 1, 2, \ldots, \ell)$ are such that the polynomial $s^\ell + k_\ell s^{\ell-1} + \cdots + k_2 s + k_1$ is Hurwitz.

Theorem 2.9 ([9]) *The system* (2.54) *subject to the control input* (2.57) *is finite-time stable at the origin with the settling time bounded by*

$$T_1(x_0) \le T_{\max} := \frac{\gamma_0}{1-\gamma_0} \frac{\lambda_{\max}(P_1)}{\lambda_{\min}(Q_1)} V^{\frac{1-\gamma_0}{\gamma_0}}(\zeta(0)), \tag{2.61}$$

where $\zeta = [x_1^{\gamma_1}, x_2^{\gamma_2}, \ldots, x_\ell^{\gamma_\ell}]^T$, $x_0 = x(0)$ *is the initial condition of the system* (2.54).

Proof Given control gains k_i, $i = 1, 2, \ldots, \ell$ such that the matrix A_1 is Hurwitz, the function $V_1(x) = x^T P_1 x$ is a Lyapunov function for linear system $\dot{x}(t) = A_1 x(t)$ and its full time derivative satisfies

$$\dot{V}_1(x) = x^T (A_1^T P_1 + P_1 A_1) x = -x^T Q_1 x < 0. \tag{2.62}$$

Using the fact that $V_1(x) \le \lambda_{\max}(P_1)\|x\|^2$ and $\dot{V}_1(x) \le -\lambda_{\min}(Q_1)\|x\|^2$ for (2.62), one gets

$$\dot{V}_1(x) \le -\frac{\lambda_{\min}(Q_1)}{\lambda_{\max}(P_1)} V_1(x) < -\frac{\lambda_{\min}(Q_1) - \delta}{\lambda_{\max}(P_1)} V_1(x) \tag{2.63}$$

for an arbitrarily small $\delta > 0$.

Since the right-hand side of system (2.60) with respect to γ_0 is continuous, $\dot{V}_1(\zeta) < 0$ also holds for the function $V_1(\zeta) = \zeta^T P_1 \zeta$ calculated along trajectories of system (2.60) if γ_0 is selected on the interval $(1 - \varepsilon, 1)$ for a sufficiently small $\varepsilon > 0$. Therefore, $V_1(\zeta) = \zeta^T P_1 \zeta$ is a Lyapunov function for system (2.60), which, in turn, is asymptotically stable at the origin. Furthermore, it can be verified that the right-hand side of system (2.60) is a homogeneous vector field of degree $\kappa = (\gamma_0 - 1)/\gamma_0 < 0$ with respect to dilations $r_i = 1/\gamma_i$, $i = 1, 2, \ldots, \ell$. Thus, it follows from Lemma 2.2 that system (2.60) is finite-time stable at the origin.

It can be verified that the Lyapunov function $V_1(\zeta) = \zeta^T P_1 \zeta$ is homogeneous in $x = [x_1, x_2, \ldots, x_\ell]^T$ of degree $l = 1 > \max\{-\kappa, 0\}$, if γ_0 is sufficiently close to 1, and $\dot{V}_1(\zeta)$ is homogeneous in $x = [x_1, x_2, \ldots, x_\ell]^T$ of degree $l + \kappa = 1 + \kappa > 0$, with respect to the same dilations r_i, $i = 1, 2, \ldots, \ell$. Using Lemma 2.3 by letting $V_2 = \dot{V}_1(\zeta)$, one obtains

$$\dot{V}_1(\zeta) \leq -cV_1^{1+\kappa}(\zeta) = -cV_1^{\frac{2\gamma_0-1}{\gamma_0}}(\zeta), \tag{2.64}$$

where c is a positive constant and $(2\gamma_0 - 1)/\gamma_0 < 1$.

On the other hand, using (2.63) and continuity of the right-hand side of system (2.60) with respect to γ_0, the inequality

$$\dot{V}_1(\zeta) < -\frac{\lambda_{\min}(Q_1) - \delta}{\lambda_{\max}(P_1)} V_1^{\frac{2\gamma_0-1}{\gamma_0}}(\zeta)$$

also holds for $V_1(\zeta)$, if γ_0 is selected on the interval $(1 - \varepsilon, 1)$ for a sufficiently small $\varepsilon > 0$ and $\dot{V}_1(\zeta)$ is calculated along trajectories of system (2.60). Since $\delta > 0$ is an arbitrarily small constant, one gets

$$\dot{V}_1(\zeta) \leq -\frac{\lambda_{\min}(Q_1)}{\lambda_{\max}(P_1)} V_1^{\frac{2\gamma_0-1}{\gamma_0}}(\zeta). \tag{2.65}$$

Therefore, $c = \lambda_{\min}(Q_1)/\lambda_{\max}(P_1)$ can be assigned in (2.64). Integrating both sides of (2.65) via separation of variables yields the settling time estimate $T_{\max}$ in (2.61). This completes the proof. □

Consider the system (2.54) subject to the control input (2.58). The resultant closed-loop system takes the form of

$$\dot{x}(t) = Ax(t) - B\Big(\sum_{i=1}^{\ell} \bar{k}_i |x_i|^{\bar{\gamma}_i} \mathrm{sign}(x_i)\Big). \tag{2.66}$$

Define the Lyapunov function candidate for system (2.66) as $V_2(\xi) = \xi^T P_2 \xi$ with $\xi = [x_1^{\bar{\gamma}_1}, x_2^{\bar{\gamma}_2}, \ldots, x_\ell^{\bar{\gamma}_\ell}]^T$ and $P_2 = P_2^T > 0$ is the solution of algebraic Lyapunov function $P_2 A_2 + A_2^T P_2 = -Q_2$ for arbitrary $Q_2 > 0$ and A_2 is in the controllable canonical form:

$$A_1 = \begin{bmatrix} 0 & 1 & \cdots & 0 \\ \vdots & \vdots & \ddots & \vdots \\ 0 & 0 & \cdots & 1 \\ -\bar{k}_1 & -\bar{k}_2 & \cdots & -\bar{k}_\ell \end{bmatrix},$$

where $\bar{k}_i > 0$, $(i = 1, 2, \ldots, \ell)$ are such that the polynomial $s^\ell + \bar{k}_\ell s^{\ell-1} + \cdots + \bar{k}_2 s + \bar{k}_1$ is Hurwitz.

Theorem 2.10 ([9]) *Consider the system* (2.54) *subject to the control input* (2.58). *For any given number* $\Upsilon > 0$ *such that* $V_2(\xi(0)) > \Upsilon$, *the trajectory of the closed-loop system* (2.66) *reaches the set* $\mathscr{S} = \{x : \|\xi\|^2 \leq \frac{\Upsilon}{\lambda_{\min}(P_2)}\}$ *within a time* $T_2 = \frac{\bar{\gamma}_0}{\bar{\gamma}_0 - 1}\frac{\lambda_{\max}(P_2)}{\lambda_{\min}(Q_2)}\Upsilon^{-(\bar{\gamma}_0-1)/\bar{\gamma}_0}$.

Proof Given control gains $\bar{k}_i, i = 1, 2, \ldots, \ell$ such that the matrix A_2 is Hurwitz, the linear system $\dot{x}(t) = A_2 x(t)$ is asymptotically stable. $V_2 = x^T P_2 x$ is a Lyapunov function for this linear system and its full time derivative of V_2 satisfies

$$\dot{V}_2 = -x^T Q_2 x \leq -\frac{\lambda_{\min}(Q_2)}{\lambda_{\max}(P_2)}V_2(x) < -\frac{\lambda_{\min}(Q_2) - \delta}{\lambda_{\max}(P_2)}V_2(x), \tag{2.67}$$

where $\delta > 0$ is an arbitrarily small constant.

In view of continuity of the right-hand side of system (2.66) with respect to $\bar{\gamma}_0$, $V_2(\xi)$ is a Lyapunov function for system (2.66) and $\dot{V}_2(\xi) < 0$ holds, if $\bar{\gamma}_0$ is selected on the interval $(1, 1+\varepsilon)$, for a sufficiently small $\varepsilon > 0$. Hence, system (2.66) is asymptotically stable. Note that the right-hand side of system (2.66) is a homogeneous vector field of degree $\kappa = (\bar{\gamma}_0 - 1)/\bar{\gamma}_0 > 0$ with respect to dilations $r_i = 1/\bar{\gamma}_i, i = 1, 2, \ldots, \ell$.

Similarly, it can be verified that the Lyapunov function $V_2(\xi)$ is homogeneous in $x = [x_1, x_2, \ldots, x_\ell]^T$ of degree $l = 1 > \max\{-\kappa, 0\}$, if $\bar{\gamma}_0$ is sufficiently close to 1, and $\dot{V}_1(\zeta)$ is homogeneous in $x = [x_1, x_2, \ldots, x_\ell]^T$ of degree $l + \kappa = 1 + \kappa > 0$, with respect to the same dilations $r_i, i = 1, 2, \ldots, \ell$. Using Lemma 2.3 and following the same argument as in the proof of Theorem 2.9, an inequality similar to (2.67) can be obtained as follows:

$$\dot{V}_2(\zeta) \leq -\frac{\lambda_{\min}(Q_2)}{\lambda_{\max}(P_2)}V_2^{\frac{2\bar{\gamma}_0-1}{\bar{\gamma}_0}}(\xi). \tag{2.68}$$

Using separation of variables and integrating both sides of (2.68) yields

$$V_2^{-\frac{\bar{\gamma}_0-1}{\bar{\gamma}_0}}(\xi(t)) - V_2^{-\frac{\bar{\gamma}_0-1}{\bar{\gamma}_0}}(\xi(0)) = \frac{\bar{\gamma}_0 - 1}{\bar{\gamma}_0}\frac{\lambda_{\min}(Q_2)}{\lambda_{\max}(P_2)}t.$$

Thus, for a given number $\Upsilon > 0$ such that $V_2(\xi(0)) > \Upsilon$, $V_2(\xi)$ decreases and reaches the value Υ at a time that can be computed as

$$\begin{aligned} t &= \frac{\bar{\gamma}_0}{\bar{\gamma}_0 - 1}\frac{\lambda_{\max}(P_2)}{\lambda_{\min}(Q_2)}\left(\Upsilon^{-\frac{\bar{\gamma}_0-1}{\bar{\gamma}_0}} - V_2^{-\frac{\bar{\gamma}_0-1}{\bar{\gamma}_0}}(\xi(0))\right) \\ &\leq \frac{\bar{\gamma}_0}{\bar{\gamma}_0 - 1}\frac{\lambda_{\max}(P_2)}{\lambda_{\min}(Q_2)}\Upsilon^{-\frac{\bar{\gamma}_0-1}{\bar{\gamma}_0}} \triangleq T_2. \end{aligned}$$

Since $\|\xi\|^2 \leq \frac{V_2(\xi)}{\lambda_{\min}(P_2)}$, the system trajectories of (2.66) reach and stay in the set $\mathscr{S} = \{x : \|\xi\|^2 \leq \frac{\Upsilon}{\lambda_{\min}(P_2)}\}$ for $t > T_2$. □

With the above preparations, we are now in a position to give the following result on fixed-time stability of system (2.54) with the control input (2.56).

Theorem 2.11 ([9]) *System (2.54) can be stabilized at the origin in fixed time by using control input* (2.56). *Moreover, the convergence time is bounded by*

$$T \leq T_{\max} := \frac{\gamma_0 \lambda_{\max}(P_1)}{(1-\gamma_0)\lambda_{\min}(Q_1)} \lambda_{\max}^{\frac{1-\gamma_0}{\gamma_0}}(P_1) + \frac{\bar{\gamma}_0 \lambda_{\max}(P_2)}{(\bar{\gamma}_0 - 1)\lambda_{\min}(Q_2)} \lambda_{\max}^{\frac{\bar{\gamma}_0-1}{\bar{\gamma}_0}}(P_2). \tag{2.69}$$

Proof Consider any given $\Upsilon > 0$ such that $V_2(\xi(0)) > \Upsilon$. Differentiating $V_2(\xi)$ against the time along trajectories of system (2.54) with the control input (2.56), one obtains

$$\begin{aligned}\dot{V}_2(\xi) &= \frac{\partial V_2}{\partial x}(Ax + Bw) + \frac{\partial V_2}{\partial x} Bv \\ &\leq \dot{V}_2(\xi) - \frac{\partial V_2}{\partial x_n}\Big(\sum_{i=1}^{\ell} k_i |x_i|^{\gamma_i} \operatorname{sign}(x_i)\Big) \\ &\leq \dot{V}_2(\xi) \leq -\frac{\lambda_{\min}(Q_2)}{\lambda_{\max}(P_2)} V_2^{\frac{2\bar{\gamma}_0-1}{\bar{\gamma}_0}}(\xi).\end{aligned}$$

Without loss of generality, Theorem 2.10 ensures that the system trajectories enter the set $\mathscr{S} = \{x : \|\xi\|^2 \leq \frac{\Upsilon}{\lambda_{\min}(P_2)} \leq 1\}$ for $t \geq T_2 = \frac{\bar{\gamma}_0}{\bar{\gamma}_0 - 1} \frac{\lambda_{\max}(P_2)}{\lambda_{\min}(Q_2)} \lambda_{\min}^{-(\bar{\gamma}_0-1)/\bar{\gamma}_0}(P_2)$.

Likewise, differentiating $V_1(\zeta)$ versus time along trajectories of system (2.54) with the control input (2.56) yields

$$\begin{aligned}\dot{V}_1(\zeta) &= \frac{\partial V_1}{\partial x}(Ax + Bv) + \frac{\partial V_1}{\partial x} Bw \\ &\leq \dot{V}_1(\zeta) - \frac{\partial V_1}{\partial x_n}\Big(\sum_{i=1}^{\ell} \bar{k}_i |x_i|^{\bar{\gamma}_i} \operatorname{sign}(x_i)\Big) \\ &\leq \dot{V}_1(\zeta) \leq -\frac{\lambda_{\min}(Q_1)}{\lambda_{\max}(P_1)} V_1^{\frac{2\gamma_0-1}{\gamma_0}}(\xi).\end{aligned}$$

Theorem 2.9 ensures that system (2.54) with the control input (2.56) is globally finite-time stable at the origin with the settling time T_1 given in (2.61). Without loss of generality, for all $t \geq T_2$, $\|\zeta\| = 1$ is applied for $V_1(\zeta(T_2)) \leq \lambda_{\max}(P_1)\|\zeta\|^2 = \lambda_{\max}(P_1)$. Thus, the settling time upper bound (2.69) can be derived. This completes the proof. □

A similar result is derived in [10] based on a bi-limit homogeneity technique [11] instead of the Lyapunov method. However, the settling time estimate cannot be explicitly provided by homogeneity formulation and only the existence of a globally bounded settling time can be guaranteed. In return, for the parameter ϵ used in

Theorem 2.11, an explicit condition $\epsilon \in (\frac{\ell-2}{\ell-1}, 1)$ that depends upon the order of the system is derived in [10] to fulfill the homogeneity.

2.4 Preliminaries on Graph Theory

The interaction topology among N agents is denoted by a digraph $\mathscr{G} = (\nu, \mathscr{E}, \mathscr{A})$, where $\nu = \{1, 2, \ldots, N\}$ is an index set, $\mathscr{E} = \{(i, j) \mid i, j \in \nu\}$ is an edge set consisting of interaction links, and $\mathscr{A} = [a_{ij}]_{N\times N}$ is a weighted adjacency matrix with weights $a_{ij} > 0$ if $(i, j) \in \mathscr{E}$ and $a_{ij} = 0$ otherwise. It is common to assume that $(i, i) \notin \mathscr{E}$ or $a_{ii} = 0$, and thus no self-loops exist. A directed edge (i, j) represents that agent j can receive information from agent i, but not vice versa for undirected edges. Denote by $\mathscr{N}_j = \{i \in \nu \mid (i, j) \in \mathscr{E}\}$ the neighbor set of agent j. Denote by $\mathscr{L} = [l_{ij}]_{N\times N}$ the Laplacian matrix of $\mathscr{G}$ with $l_{ij} = -a_{ij}$ for $i \neq j$, and $l_{ii} = \sum_{j=1}^{N} a_{ij}$. A directed path of $\mathscr{G}$ is denoted by a sequence of directed edges. $\mathscr{G}$ has a directed spanning tree if there exists at least one agent that has a directed path to any other agent. $\mathscr{G}$ is called strongly connected if there exists a directed path for any two agents. For an undirected graph, $a_{ij} = a_{ji}$.

Let $\bar{\mathscr{G}}$ be an interaction graph among N followers and a leader with index 0. The leader only sends out information and receives no information from any other agents. When the ith follower has information access of the leader, a notation, b_i, is set to be a positive constant, and to be zero otherwise. Then, a matrix, $\mathscr{M}$, is defined as $\mathscr{M} = \mathscr{L} + \mathscr{B}$, where $\mathscr{B} = \text{diag}(b_1, b_2, \ldots, b_N)$. Note that a directed graph contains a directed spanning tree if and only if there exists at least one agent that has a directed path to any other agent.

Lemma 2.4 *([12]) The matrix $\mathscr{L}$ has the following properties:*

1. *0 is an eigenvalue of $\mathscr{L}$ and $\mathbf{1}_N$ is the associated eigenvector;*
2. *$x^T\mathscr{L}x = \frac{1}{2}\sum_{i=1}^{N}\sum_{j=1}^{N} a_{ij}(x_j - x_i)^2$, and the positive semi-definiteness of $\mathscr{L}$ implies that all eigenvalues of $\mathscr{L}$ are real and nonnegative;*
3. *If $\mathscr{G}$ is connected, the second smallest eigenvalue of $\mathscr{L}$, which is denoted by $\lambda_2(\mathscr{L})$ and called the algebraic connectivity of $\mathscr{G}$ is greater than zero;*
4. *The algebraic connectivity of $\mathscr{G}$ is equal to $\min_{x\neq 0, \mathbf{1}_N^T x=0} \frac{x^T\mathscr{L}x}{x^Tx}$, and therefore, if $\mathbf{1}_N^T x = 0$, $x^T\mathscr{L}x \geq \lambda_2(\mathscr{L})x^Tx$.*

Lemma 2.5 *([13]) If $\bar{\mathscr{G}}$ contains a directed spanning tree with the leader as the root, then $\mathscr{M}$ is nonsingular. Furthermore, define $[p_1, \ldots, p_N]^T = \mathscr{M}^{-T}\mathbf{1}_N$, $P = diag(p_1, \ldots, p_N)$, $Q = P\mathscr{M} + \mathscr{M}^T P$, then P and Q are both positive definite.*

2.5 Conclusions

The motivation and insight behind some developments of a fixed-time stability concept are illustrated by introducing three interesting examples. Formal definitions on fixed-time stability are given. The key feature, different from finite-time stability, is that the settling time is upper bounded by a constant independent of system initial conditions. Some important fixed-time stability theorems are presented, which establish the relationship between the regularity properties of the Lyapunov function and the convergence time. Finally, two fixed-time stabilization control approaches, i.e., a terminal sliding mode approach and a homogeneity and Lyapunov approach are introduced. These two approaches are fundamental approaches in fixed-time cooperative control.

2.6 Notes

The benchmark work [3] links the regularity properties of Lyapunov functions and finite-time stability. Thereafter, a number of finite-time stabilization control schemes are developed based on a Lyapunov function method [14], a high-order sliding mode method [15, 16], and a homogeneity method [8]. Fixed-time stability was first found in [11] and was formally defined in [4]. The main methods for analysis of fixed-time stability include a Lyapunov function method [17] and a homogeneity method [11]. Furthermore, an implicit Lyapunov function (ILF) method, which defines Lyapunov functions implicitly as solutions to an algebraic equation, is developed in [18] for finite-time and fixed-time stability analysis of nonlinear systems.

References

1. A. Filippov, *Differential Equations With Discontinuous Right-Hand Side* (Kluwer Academic, Dordrecht, 1988)
2. H.K. Khalil, *Nonlinear Systems* (Prentice Hall, New Jersey, 2002)
3. S. Bhat, D. Bernstein, Finite-time stability of continuous autonomous systems. SIAM J. Control Optim. **38**(3), 751–766 (2000)
4. A. Polyakov, Nonlinear feedback design for fixed-time stabilization of linear control systems. IEEE Trans. Automa. Control **57**(8), 2106–2110 (2012)
5. Z. Zuo, L. Tie, A new class of finite-time nonlinear consensus protocols for multi-agent systems. Int. J. Control **87**(2), 363–370 (2014)
6. Z. Zuo, L. Tie, Distributed robust finite-time nonlinear consensus protocols for multi-agent systems. Int. J. Syst. Sci. **47**(6), 1366–1375 (2016)
7. S. Parsegov, A. Polyakov, P. Shcherbakov, *Nonlinear fixed-time control protocol for uniform allocation of agents on a segment, In Proceedings of the IEEE Conference on Decision and Control* (Hawaii, USA, 2013), pp. 7732–7737
8. S. Bhat, D. Bernstein, Geometric homogeneity with applications to finite-time stability. Math. Control Signals Syst. **17**(2), 101–127 (2005)

9. M. Basin, Y. Shtessel, F. Aldukali, Continuous finite- and fixed-time high-order regulators. J. Frankl. Inst. **253**(18), 5001–5012 (2016)
10. B. Tian, H. Lu, Z. Zuo, H. Wang, Fixed-time stabilization of high-order integrator systems with mismatched disturbances. Nonlinear Dyn. **94**(4), 2889–2899 (2018)
11. A.A. Vincent Andrieu, L. Praly, Homogeneous approximation, recursive observer design and output feedback, SIAM J. Control Optim. **47**(4), pp. 1814–1850 (2008)
12. R. Olfati-Saber, R.M. Murray, Consensus problems in networks of agents with switching topology and time-delays. IEEE Trans. Autom. Control **49**(9), 1520–1533 (2004)
13. J. Fu, J. Wang, Finite-time consensus for multi-agent systems with globally bounded convergence time under directed communication graphs. Int. J. Control **90**(9), 1807–1817 (2017)
14. A. Polyakov, A. Poznyak, Lyapunov function design for finite-time convergence analysis: "twisting" controller for second-order sliding mode realization. Automatica **45**(2), 444–448 (2009)
15. A. Levant, Homogeneity approach to high order sliding mode design. Automatica **41**(5), 823–830 (2005)
16. A. Levant, Quasi-continuous high-order sliding mode controllers. IEEE Trans. Autom. Control **50**(11), 1812–1816 (2005)
17. A. Polyakov, L. Fridman, Stability notions and Lyapunov functions for sliding mode control systems. J. Frankl. Inst. **351**(4), 1831–1865 (2014)
18. A. Polyakov, D. Efimov, W. Perruquetti, Finite-time and fixed-time stabilization: implicit Lyapunov function approach. Automatica **51**(2), 332–340 (2015)

Chapter 3
Fixed-Time Cooperative Control for First-Order Multi-Agent Systems

This chapter addresses a fixed-time consensus problem for a multi-agent system in networks with undirected topology. A class of global continuous time-invariant consensus protocols is constructed for single-integrator agents. It is shown that the settling time of the proposed fixed-time consensus protocols is upper bounded for arbitrary initial conditions. Therefore, the convergence time can be designed and estimated off-line for a given undirected information flow and a group of agents.

3.1 Problem Formulation

Consider a group of N continuous-time agents with dynamics in the form of

$$\dot{x}_i(t) = u_i(t), \quad i \in \Gamma = \{1, 2, \ldots, N\}, \tag{3.1}$$

where $x_i \in \mathbb{R}$ is the state of agent $i \in \Gamma$, and $u_i \in \mathbb{R}$ is the controller, or called a *protocol*, to be designed. With a given protocol u_i, $i \in \Gamma$, the corresponding closed-loop system of system (3.1) is said to achieve *asymptotic consensus*, if, for $\forall x_i(0)$ and $\forall i, j \in \Gamma$, $|x_i(t) - x_j(t)| \to 0$ as $t \to \infty$.

Definition 3.1 For system (3.1), it is said to achieve fixed-time consensus if there exists a suitable protocol $u_i(t)$ such that

$$\begin{cases} \lim\limits_{t \to T} |x_i(t) - x_j(t)| = 0; \\ |x_i(t) - x_j(t)| = 0, \quad \forall t \geq T, \end{cases}$$

for all $i, j \in \Gamma$, where T is upper bounded by $T_{\max}$ that is independent of initial states.

Z. Zuo et al., *Fixed-Time Cooperative Control of Multi-Agent Systems*,
https://doi.org/10.1007/978-3-030-20279-8_3

If the final agreement state satisfies

$$\lim_{t\to\infty} x_i(t) = \frac{1}{N}\sum_{k=1}^{N} x_k(0), \quad \forall i \in \Gamma, \tag{3.2}$$

then the system is said to achieve *average consensus*.

Stack the states of N agents as a vector $x = [x_1, x_2, \ldots, x_N]^T$, and let $\text{span}(\mathbf{1}_N) = \{\xi \in \mathbb{R}^N : \xi = r\mathbf{1}_N, r \in \mathbb{R}\}$.

The dynamics of the leader is described by

$$\dot{x}_0(t) = u_0(t), \tag{3.3}$$

where $x_0(t) \in \mathbb{R}$ is the leader's state, and $u_0(t) \in \mathbb{R}$ is the control input for the leader.

We now give the definition of fixed-time consensus tracking for the system described by (3.1) and (3.3).

Definition 3.2 For the system described by (3.1) and (3.3), it is said to achieve fixed-time consensus tracking if there exists a suitable protocol $u_i(t)$ such that

$$\begin{cases} \lim_{t\to T} |x_i(t) - x_0(t)| = 0; \\ |x_i(t) - x_0(t)| = 0, \quad \forall t \geq T \end{cases}$$

for all $i \in \Gamma$, where T is upper bounded by $T_{\max}$ that is independent of initial states.

The following two assumptions are made.

Assumption 3.1.1 The control input of the leader is unknown to its followers, but it is bounded by a known positive constant $u_0^{\max}$, i.e.,

$$|u_0| \leq u_0^{\max} < \infty. \tag{3.4}$$

Assumption 3.1.2 The undirected graph $\mathcal{G}$ is connected, and at least one follower can directly obtain the information of the leader, i.e., at least one $b_i > 0$. □

To streamline technical proofs of main results, the following three lemmas are needed.

Lemma 3.1 *Let* $\xi_1, \xi_2, \ldots, \xi_N \geq 0$ *and* $0 < p \leq 1$. *Then*

$$\sum_{i=1}^{N} \xi_i^p \geq \Big(\sum_{i=1}^{N} \xi_i\Big)^p. \tag{3.5}$$

Proof Noting that $x^p \geq x$ for $\forall x$, $p \in (0, 1]$, one obtains

$$\frac{\sum_{i=1}^N \xi_i^p}{(\sum_{i=1}^N \xi_i)^p} = \sum_{i=1}^N \left(\frac{\xi_i}{\sum_{i=1}^N \xi_i}\right)^p \geq \sum_{i=1}^N \frac{\xi_i}{\sum_{i=1}^N \xi_i} = 1,$$

which is equivalent to (3.5). □

Lemma 3.2 *Let $\xi_1, \xi_2, \ldots, \xi_N \geq 0$ and $p > 1$. Then*

$$\sum_{i=1}^N \xi_i^p \geq N^{1-p}\Big(\sum_{i=1}^N \xi_i\Big)^p. \tag{3.6}$$

Proof Based on *Hölder inequality*

$$\sum_{i=1}^N \xi_i \eta_i \leq \Big(\sum_{i=1}^N \xi_i^p\Big)^{\frac{1}{p}}\Big(\sum_{j=1}^N \eta_j^q\Big)^{\frac{1}{q}}$$

for $\forall \xi_i, \eta_i > 0,\ i \in \Gamma$ and $\forall p, q \in (1, \infty)$ satisfying $1/p + 1/q = 1$, (3.6) is a straightforward result by letting $[\eta_1, \eta_2, \ldots, \eta_N]^T = \mathbf{1}_N^T$. □

Lemma 3.3 *For any vector $x \in \mathbb{R}^\ell$, if $p > r > 0$, where p and r are scalar constants, then the following inequality holds*

$$\|x\|_p \leq \|x\|_r \leq \ell^{\frac{1}{r}-\frac{1}{p}}\|x\|_p. \tag{3.7}$$

Proof Since $\big(\sum_{i=1}^\ell |x_i|^p\big)^{1/p} \leq \big(\sum_{i=1}^\ell |x_i|^r\big)^{1/r}$ [1], the first inequality holds directly. Then applying Hölder's inequality, we get

$$\begin{aligned}\sum_{i=1}^\ell |x_i|^r = \sum_{i=1}^\ell 1 \cdot |x_i|^r &\leq \Big(\sum_{i=1}^\ell 1^{\frac{p}{p-r}}\Big)^{1-\frac{r}{p}}\Big(\sum_{i=1}^\ell (|x_i|^r)^{\frac{p}{r}}\Big)^{\frac{r}{p}} \\ &= \ell^{1-\frac{r}{p}}\Big(\sum_{i=1}^\ell |x_i|^p\Big)^{\frac{r}{p}}.\end{aligned} \tag{3.8}$$

Therefore,

$$\|x\|_r = \Big(\sum_{i=1}^\ell |x_i|^r\Big)^{\frac{1}{r}} \leq \ell^{\frac{1}{r}-\frac{1}{p}}\Big(\sum_{i=1}^\ell |x_i|^p\Big)^{\frac{1}{p}} = \ell^{\frac{1}{r}-\frac{1}{p}}\|x\|_p. \tag{3.9}$$

So the second inequality of (3.7) holds, which completes the proof. □

3.2 Fixed-Time Consensus

In this section, we develop nonlinear fixed-time continuous protocols for a multi-agent system (3.1). The protocols are designed in the following form:

$$u_i = \alpha\Big(\sum_{j\in\mathcal{N}_j} a_{ij}(x_j - x_i)\Big)^{2-\frac{p}{q}} + \beta\Big(\sum_{j\in\mathcal{N}_j} a_{ij}(x_j - x_i)\Big)^{\frac{p}{q}}; \tag{3.10}$$

$$u_i = \alpha \sum_{j\in\mathcal{N}_j} a_{ij}(x_j - x_i)^{2-\frac{p}{q}} + \beta \sum_{j\in\mathcal{N}_j} a_{ij}(x_j - x_i)^{\frac{p}{q}}, \tag{3.11}$$

where $\alpha, \beta > 0$, p, q are both positive odd integers satisfying $p < q$.

It is worthwhile mentioning in advance that the protocol (3.11) solves fixed-time *average* consensus problem while the protocol (3.10) only achieves fixed-time consensus, which will be clearly shown hereinafter.

Remark 3.1 Protocols (3.10) and (3.11) are both continuous functions w.r.t state variables provided that the interaction topology is time invariant. If setting $p = q = 1$, the two protocols are degenerated into the typical linear consensus protocol in [2] which achieves asymptotical agreement. Note that protocols (3.10) and (3.11) have a faster convergence rate than the linear protocol [2] and the combined linear–nonlinear protocol [3].

Now, we are in a position to state and establish the main results in this chapter, which are summarized in the following theorems.

Theorem 3.1 *Suppose that the undirected graph $\mathcal{G}$ with $\mathcal{A}$ of system (3.1) is connected. Then the distributed consensus protocol (3.10) achieves fixed-time stability.*

Proof Consider the following positive semi-definite function

$$V(x(t)) = \frac{1}{2}x^T\mathcal{L}x = \frac{1}{4}\sum_{i=1}^{N}\sum_{j=1}^{N} a_{ij}\left(x_j(t) - x_i(t)\right)^2.$$

Since $\mathcal{G}$ is connected, zero is a simple eigenvalue of $\mathcal{L}$ which implies that $V(x(t)) = 0$ if and only if $x(t) \in \text{span}\{\mathbf{1}_N\}$ by Lemma 2.4. The symmetry of $\mathcal{A}$ guarantees

$$\frac{\partial V(x)}{\partial x_i} = -\sum_{j=1}^{N} a_{ij}(x_j - x_i).$$

Thus, the derivative of $V(x)$ with respect to time is

$$
\begin{aligned}
\dot{V}(x) &= \sum_{i=1}^{N} \frac{\partial V(x)}{\partial x_i} \dot{x}_i \\
&= -\alpha \sum_{i=1}^{N} \Big(\sum_{j\in\mathcal{N}_j} a_{ij}(x_j - x_i) \Big)^{\frac{3q-p}{q}} - \beta \sum_{i=1}^{N} \Big(\sum_{j\in\mathcal{N}_j} a_{ij}(x_j - x_i) \Big)^{\frac{p+q}{q}} \\
&= -\alpha \sum_{i=1}^{N} \Big(\Big(\sum_{j\in\mathcal{N}_j} a_{ij}(x_j - x_i) \Big)^2 \Big)^{\frac{3q-p}{2q}} - \beta \sum_{i=1}^{N} \Big(\Big(\sum_{j\in\mathcal{N}_j} a_{ij}(x_j - x_i) \Big)^2 \Big)^{\frac{p+q}{2q}} \\
&\le -\alpha N^{\frac{p-q}{2q}} \Big(\sum_{i=1}^{N} \Big(\sum_{j\in\mathcal{N}_j} a_{ij}(x_j - x_i) \Big)^2 \Big)^{\frac{3q-p}{2q}} \\
&\quad - \beta \Big(\sum_{i=1}^{N} \Big(\sum_{j\in\mathcal{N}_j} a_{ij}(x_j - x_i) \Big)^2 \Big)^{\frac{p+q}{2q}},
\end{aligned} \tag{3.12}
$$

where *Lemmas* 3.1 and 3.2 are applied in view of $(3q - p)/(2q) \in (1, \infty)$ and $(p + q)/(2q) \in (0, 1]$. The positive semi-definiteness of $\mathcal{L}$ ensures that there exists $Q \in \mathbb{R}^{N\times N}$ such that $\mathcal{L} = Q^T Q$. By Lemma 2.4, for $V(x) \neq 0$, one obtains

$$
\begin{aligned}
\frac{\sum_{i=1}^{N} \Big(\sum_{j\in\mathcal{N}_j} a_{ij}(x_j - x_i) \Big)^2}{V(x)} &= \frac{x^T \mathcal{L}^T \mathcal{L} x}{\frac{1}{2} x^T \mathcal{L} x} \\
&= \frac{2x^T Q^T Q Q^T Q x}{x^T Q^T Q x} = \frac{2x^T Q^T \mathcal{L}^T Q x}{x^T Q^T Q x} \ge 2\lambda_2(\mathcal{L}),
\end{aligned} \tag{3.13}
$$

where $\lambda_2(\mathcal{L}) > 0$ by Lemma 2.14. With (3.13) being applied for (3.12), one obtains

$$
\begin{aligned}
\dot{V} &\le -\alpha N^{\frac{p-q}{2q}} \left(2\lambda_2(\mathcal{L})V\right)^{\frac{3q-p}{2q}} - \beta \left(2\lambda_2(\mathcal{L})V\right)^{\frac{p+q}{2q}} \\
&= -\left[\alpha N^{\frac{p-q}{2q}} \left(2\lambda_2(\mathcal{L})V\right)^{\frac{q-p}{q}} + \beta \right] \left(2\lambda_2(\mathcal{L})V\right)^{\frac{p+q}{2q}}.
\end{aligned} \tag{3.14}
$$

If $V \neq 0$, let $y = \sqrt{2\lambda_2(\mathcal{L})V}$. Then $\dot{y} = \lambda_2(\mathcal{L})\dot{V}/\sqrt{2\lambda_2(\mathcal{L})V}$. From (3.14), one gets that

$$
\dot{y}(t) \le -\alpha N^{\frac{p-q}{2q}} \lambda_2(\mathcal{L}) y^{\frac{2q-p}{q}}(t) - \beta \lambda_2(\mathcal{L}) y^{\frac{p}{q}}(t). \tag{3.15}
$$

By using Corollary 2.1 and *Comparison Principle* of differential equations [4], one gets

$$
\lim_{t\to T} V(x) = 0,
$$

where the settling time is given by

$$
\begin{aligned}
T &= \frac{qN^{\frac{q-p}{4q}}}{\sqrt{\alpha\beta}\lambda_2(\mathscr{L})(q-p)} \arctan\left(N^{\frac{p-q}{4q}}\sqrt{\frac{\alpha}{\beta}}V(x_0)\right) \\
&\le \frac{\pi qN^{\frac{q-p}{4q}}}{2\sqrt{\alpha\beta}\lambda_2(\mathscr{L})(q-p).}
\end{aligned} \tag{3.16}
$$

Thus,

$$
\lim_{t\to T} |x_j(t) - x_i(t)| = 0, \quad \forall i, j \in \Gamma.
$$

Remark 3.2 The upper bound of the settling time T is only related to the design parameters of the protocol (3.10), the algebraic connectivity of $\mathscr{G}$, and the volume N of the multi-agent system.

Theorem 3.2 *Suppose that the undirected graph $\mathscr{G}$ with $\mathscr{A}$ of system (3.1) is connected. Then the distributed consensus protocol (3.11) achieves the fixed-time average consensus.*

Proof Since $a_{ij} = a_{ji}$ for $\forall i, j \in \Gamma$ and $(\cdot)^{2-p/q}$ and $(\cdot)^{p/q}$ are both continuous odd functions w.r.t $(\cdot)$, invoking (3.11), one gets

$$
\sum_{i=1}^{N}\dot{x}_i(t) = \sum_{i=1}^{N}u_i(t) = \alpha\sum_{i=1}^{N}\sum_{j\in\mathscr{N}_j} a_{ij}(x_j - x_i)^{2-\frac{p}{q}} + \beta\sum_{i=1}^{N}\sum_{j\in\mathscr{N}_j} a_{ij}(x_j - x_i)^{\frac{p}{q}} = 0.
$$

Let $x^* = (1/N)\sum_{i=1}^{N} x_i(t)$ and $\delta_i(t) = x_i(t) - x^*$. It follows from $\sum_{i=1}^{N}\dot{x}_i(t) = 0$ that x^* is time invariant and therefore $\dot{\delta}_i(t) = \dot{x}_i(t)$. Then the group disagreement vector [5] can be written as $\delta(t) = [\delta_1(t), \delta_2(t), \ldots, \delta_N(t)]^T$. Consider the following candidate Lyapunov function:

$$
V(\delta(t)) = \frac{1}{2}\sum_{i=1}^{N}\delta_i^2(t).
$$

Differentiating $V(\delta(t))$ with respect to time and applying *Lemmas* 3.1 and 3.2, we have

$$
\begin{aligned}
\dot{V}(\delta(t)) &= \sum_{i=1}^{N}\delta_i\dot{\delta} = \alpha\sum_{i=1}^{N}\delta_i\sum_{j=1}^{N}a_{ij}(\delta_j - \delta_i)^{2-\frac{p}{q}} + \beta\sum_{i=1}^{N}\delta_i\sum_{j=1}^{N}a_{ij}(\delta_j - \delta_i)^{\frac{p}{q}} \\
&= \frac{1}{2}\alpha\sum_{i=1}^{N}\sum_{j=1}^{N}a_{ij}(\delta_i - \delta_j)(\delta_j - \delta_i)^{2-\frac{p}{q}} + \frac{1}{2}\beta\sum_{i=1}^{N}\sum_{j=1}^{N}a_{ij}(\delta_i - \delta_j)(\delta_j - \delta_i)^{\frac{p}{q}} \\
&= -\frac{1}{2}\alpha\sum_{i=1}^{N}\sum_{j=1}^{N}a_{ij}(\delta_j - \delta_i)^{\frac{3q-p}{q}} - \frac{1}{2}\beta\sum_{i=1}^{N}\sum_{j=1}^{N}a_{ij}(\delta_j - \delta_i)^{\frac{p+q}{q}}
\end{aligned}
$$

$$
\begin{aligned}
&= -\frac{1}{2}\alpha \sum_{i=1}^{N}\sum_{j=1}^{N}\left[a_{ij}^{\frac{2q}{3q-p}}(\delta_j-\delta_i)^2\right]^{\frac{3q-p}{2q}} - \frac{1}{2}\beta \sum_{i=1}^{N}\sum_{j=1}^{N}\left[a_{ij}^{\frac{2q}{p+q}}(\delta_j-\delta_i)^2\right]^{\frac{p+q}{2q}} \\
&\leq -\frac{1}{2}\alpha N^{\frac{p-q}{q}}\left[\sum_{i=1}^{N}\sum_{j=1}^{N} a_{ij}^{\frac{2q}{3q-p}}(\delta_j-\delta_i)^2\right]^{\frac{3q-p}{2q}} \\
&\quad - \frac{1}{2}\beta \left[\sum_{i=1}^{N}\sum_{j=1}^{N} a_{ij}^{\frac{2q}{p+q}}(\delta_j-\delta_i)^2\right]^{\frac{p+q}{2q}}. \qquad (3.17)
\end{aligned}
$$

Let $G_1 = \sum_{i,j=1}^{N} b_{ij}(\delta_j-\delta_i)^2$ and $G_2 = \sum_{i,j=1}^{N} c_{ij}(\delta_j-\delta_i)^2$ with $b_{ij} \triangleq a_{ij}^{2q/(3q-p)}$ and $c_{ij} \triangleq a_{ij}^{2q/(p+q)}$, for $\forall i, j \in \Gamma$, respectively. Let $B = [b_{ij}]_{N\times N}$ and $C = [c_{ij}]_{N\times N}$. Then one gets that $G_B(\delta) = 2\delta^T \mathscr{L}_B \delta$ and $G_C(\delta) = 2\delta^T \mathscr{L}_C \delta$, where $\mathscr{L}_B$ and $\mathscr{L}_C$ are, respectively, the graph Laplacians of $\mathscr{G}(B)$ and $\mathscr{G}(C)$. It follows from the definitions of b_{ij} and c_{ij} that both $\mathscr{G}(B)$ and $\mathscr{G}(C)$ are connected if $\mathscr{G}$ is connected. Since $\mathbf{1}_N^T \delta = 0$, by Lemma 2.4, one gets

$$
\begin{cases} G_B \geq 2\lambda_2(\mathscr{L}_B)\delta^T\delta; \\ G_C \geq 2\lambda_2(\mathscr{L}_C)\delta^T\delta, \end{cases} \qquad (3.18)
$$

where $\lambda_2(\mathscr{L}_B) > 0$ and $\lambda_2(\mathscr{L}_C) > 0$. Note that $G_B/V \geq 4\lambda_2(\mathscr{L}_B)$ and $G_C/V \geq 4\lambda_2(\mathscr{L}_C)$ if $V(\delta) \neq 0$. One further obtains

$$
\begin{aligned}
\dot{V}(\delta) &\leq -\frac{1}{2}\alpha N^{\frac{p-q}{q}}\left[\frac{G_B(\delta)}{V}V\right]^{\frac{3q-p}{2q}} - \frac{1}{2}\beta\left[\frac{G_C(\delta)}{V}V\right]^{\frac{p+q}{2q}} \\
&= -\frac{1}{2}\alpha N^{\frac{p-q}{q}}\left[4\lambda_2(\mathscr{L}_B)V\right]^{\frac{3q-p}{2q}} - \frac{1}{2}\beta\left[4\lambda_2(\mathscr{L}_C)V\right]^{\frac{p+q}{2q}} \\
&\leq -\frac{1}{2}\alpha N^{\frac{p-q}{q}}\left(4\underline{\lambda}V\right)^{\frac{3q-p}{2q}} - \frac{1}{2}\beta\left(4\underline{\lambda}V\right)^{\frac{p+q}{2q}} \\
&= -\frac{1}{2}\left[\alpha N^{\frac{p-q}{q}}\left(4\underline{\lambda}V\right)^{\frac{q-p}{q}} + \beta\right]\left(4\underline{\lambda}V\right)^{\frac{p+q}{2q}}, \qquad (3.19)
\end{aligned}
$$

where $\underline{\lambda} = \min\{\lambda_2(\mathscr{L}_B), \lambda_2(\mathscr{L}_C)\} > 0$. If $V \neq 0$, let $y = \sqrt{4\underline{\lambda}V}$. Then $\dot{y} = 2\underline{\lambda}\dot{V}/\sqrt{4\underline{\lambda}V}$. It follows from (3.19) that

$$
\dot{y}(t) \leq -\alpha N^{\frac{p-q}{q}}\underline{\lambda}y^{\frac{2q-p}{q}}(t) - \beta\underline{\lambda}y^{\frac{p}{q}}(t). \qquad (3.20)
$$

By using Corollary 2.1 and *Comparison Principle*, one can obtain $\lim_{t\to T} y(t) = 0$ and therefore $\lim_{t\to T} V(t) = 0$ which, in turn, implies that

$$
\lim_{t\to T} \delta(t) = 0,
$$

where the settling time is given by

$$\begin{aligned} T &= \frac{qN^{\frac{q-p}{2q}}}{\underline{\lambda}\sqrt{\alpha\beta}(q-p)} \arctan\left(N^{\frac{p-q}{2q}}\sqrt{\frac{\alpha}{\beta}}V(\delta_0)\right) \\ &\leq \frac{\pi q N^{\frac{q-p}{2q}}}{2\underline{\lambda}\sqrt{\alpha\beta}(q-p)}. \end{aligned} \tag{3.21}$$

Since $\delta(t) \to 0$ implies that $x_i(t) \to x^*$ for $\forall i \in \Gamma$, the protocol solves the fixed-time average consensus problem. □

Remark 3.3 It is clear that the settling time is upper bounded by the design parameters, the volume N of a given agent team and the *minimum* algebraic connectivity of $\mathscr{G}(B)$ and $\mathscr{G}(C)$. Different from Theorem 3.1, the Lyapunov function used in Theorem 3.2 does not depend on the network topology, which makes it possible for stability analysis of a multi-agent system with switching topology.

3.3 Fixed-Time Consensus Tracking

In this section, we develop a fixed-time consensus tracking protocol for the multi-agent system described by (3.1) and (3.3). The protocols are designed in the following form:

$$\begin{aligned} u_i =& -\alpha\Bigg(\sum_{j=1}^{N} a_{ij}(x_i - x_j) + b_i(x_i - x_0)\Bigg)^{2-\frac{p}{q}} - \beta\Bigg(\sum_{j=1}^{N} a_{ij}(x_i - x_j) + b_i(x_i - x_0)\Bigg)^{\frac{p}{q}} \\ & - \gamma \mathrm{sign}\Bigg(\sum_{j=1}^{N} a_{ij}(x_i - x_j) + b_i(x_i - x_0)\Bigg), \end{aligned} \tag{3.22}$$

where $\gamma > 0$, and α, β, p, q are defined in Sect. 3.2.

Theorem 3.3 *Under Assumptions 3.1.1 and 3.1.2, if $\gamma \geq u_0^{\max}$, then using the distributed control protocol (3.22), the fixed-time consensus tracking is achieved for the multi-agent system described by (3.1) and (3.3). In particular, an upper bound of the settling time is given by*

$$T_{\max} = \frac{\pi q N^{\frac{q-p}{4q}}}{2\sqrt{\alpha\beta}\lambda_{\min}(\mathscr{M})(q-p)}, \tag{3.23}$$

where $\mathscr{M} = \mathscr{L} + \mathscr{B}$.

Proof Define $e_i = x_i - x_0,\ i \in \Gamma$. One gets

$$\begin{aligned}\dot{e}_i = &-\alpha\Big(\sum_{j=1}^{N} a_{ij}(e_i - e_j) + b_i(e_i - e_0)\Big)^{2-\frac{p}{q}} \\ &-\beta\Big(\sum_{j=1}^{N} a_{ij}(e_i - e_j) + b_i(e_i - e_0)\Big)^{\frac{p}{q}} \\ &-\gamma\text{sign}\Big(\sum_{j=1}^{N} a_{ij}(e_i - e_j) + b_i(e_i - e_0)\Big) - u_0. \end{aligned} \tag{3.24}$$

Denote $e = [e_1, e_2, \ldots, e_N]^T$. Then the dynamics of N agents can be written in a compact vector form

$$\dot{e} = -\alpha(\mathscr{M}e)^{2-\frac{p}{q}} - \beta(\mathscr{M}e)^{\frac{p}{q}} - \gamma\text{sign}(\mathscr{M}e) - u_0\mathbf{1}_N. \tag{3.25}$$

Construct a Lyapunov function candidate

$$V = \frac{1}{2}e^T\mathscr{M}e. \tag{3.26}$$

The time derivative of $V(e)$ is calculated as

$$\begin{aligned}\dot{V} =& e^T\mathscr{M}\dot{e} \\ =& -\alpha\|\mathscr{M}e\|_{3-\frac{p}{q}}^{3-\frac{p}{q}} - \beta\|\mathscr{M}e\|_{1+\frac{p}{q}}^{1+\frac{p}{q}} - \gamma\|\mathscr{M}e\|_1 - e^T\mathscr{M}(u_0\mathbf{1}_N) \\ \le& -\alpha\|\mathscr{M}e\|_{3-\frac{p}{q}}^{3-\frac{p}{q}} - \beta\|\mathscr{M}e\|_{1+\frac{p}{q}}^{1+\frac{p}{q}} - \gamma\|\mathscr{M}e\|_1 + u_0^{\max}\|\mathscr{M}e\|_1 \\ \le& -\alpha\|\mathscr{M}e\|_{3-\frac{p}{q}}^{3-\frac{p}{q}} - \beta\|\mathscr{M}e\|_{1+\frac{p}{q}}^{1+\frac{p}{q}}. \end{aligned} \tag{3.27}$$

Since $\mathscr{M}$ is a positive-definite matrix, there exists a unique upper triangular matrix $Q \in \mathbb{R}^{N\times N}$ satisfying $\mathscr{M} = Q^TQ$ [6]. Therefore,

$$\begin{aligned}\|\mathscr{M}e\|_2^2 =& e^T(Q^TQ)(Q^TQ)e \\ =& (e^TQ^T)(QQ^T)(Qe) \\ \ge& \lambda_{\min}(\mathscr{M})\big(Qe\big)^T\big(Qe\big) = \lambda_{\min}(\mathscr{M})e^T\mathscr{M}e, \end{aligned} \tag{3.28}$$

where the inequality holds by Theorem 3 in [7].

By using Lemma 3.3, one obtains

$$\|\mathscr{M}e\|_{1+\frac{p}{q}} \ge \|\mathscr{M}e\|_2 \ge (2\lambda_{\min}(\mathscr{M})V)^{\frac{1}{2}}, \tag{3.29}$$

Then

$$\|\mathscr{M}e\|_{1+\frac{p}{q}}^{1+\frac{p}{q}} \geq (2\lambda_{\min}(\mathscr{M})V)^{\frac{1}{2}+\frac{p}{2q}}. \tag{3.30}$$

Similarly, since $\|\mathscr{M}e\|_2 \leq (N^{1/2-1/(3-p/q)})\|\mathscr{M}e\|_{3-\frac{p}{q}}$, one gets

$$\begin{aligned}\|\mathscr{M}e\|_{3-\frac{p}{q}} &\geq \left(N^{\frac{1}{2}-\frac{1}{3-\frac{p}{q}}}\right)^{-1}\|\mathscr{M}e\|_2 \\ &\geq \left(N^{\frac{1}{2}-\frac{1}{3-\frac{p}{q}}}\right)^{-1}(2\lambda_{\min}(\mathscr{M})V)^{\frac{1}{2}}.\end{aligned} \tag{3.31}$$

Then we have that $\|\mathscr{M}e\|_{3-p/q}^{3-p/q} \geq N^{(p-q)/2q}(2\lambda_{\min}(\mathscr{M})V)^{3/2-p/(2q)}$. Therefore,

$$\dot{V} \leq -\alpha N^{\frac{p-q}{2q}}\big(2\lambda_{\min}(\mathscr{M})\big)^{\frac{3}{2}-\frac{p}{2q}}V^{\frac{3}{2}-\frac{p}{2q}} - \beta\big(2\lambda_{\min}(\mathscr{M})\big)^{\frac{1}{2}+\frac{p}{2q}}V^{\frac{1}{2}+\frac{p}{2q}}. \tag{3.32}$$

By using Corollary 2.1 and *Comparison Principle*, following a similar proof procedure of Theorem 3.1, one can obtain that V converges to zero in a fixed time, and the settling time is bounded by (3.23). Therefore, $x_i(t) = x_0(t),\ i \in \Gamma$ for $t \geq T_{\max}$, which completes the proof. □

3.4 Numerical Examples

Example 1: Fixed-time consensus. We take a 6-agent system ($N = 6$) in a network with an undirected topology illustrated in Fig. 3.1 as an example. In the simulations, nonzero $a_{ij} = 2$, $\alpha = \beta = 2$, $p = 7$, and $q = 9$ are set for each protocol. The algebraic connectivity of $\mathscr{G}$ is 0.8262, while the ones corresponding to $\mathscr{G}(B)$ and $\mathscr{G}(C)$ are 0.7709 and 0.9010, respectively. The estimated upper bounds of the settling time for protocols (3.10) and (3.11) are 4.7231s and 5.5917s by (3.16) and (3.21), respectively. Consider two initial conditions: (i) $x(0) = [-5, -2, 4, 6, 4, 5]^T$ and (ii) $x(0) = [10, -20, -3, 9, 4, -30]^T$.

The numerical results given in Figs. 3.2, 3.3, 3.4, and 3.5 show that the settling time of the two consensus protocols under different initial conditions are both about 3s, which demonstrates the correctness of the estimated bounds in *Theorems* 3.1 and 3.2. For each initial condition, the average consensus values can be computed via $x^* = \frac{1}{6}\sum_{i=1}^{6} x(0)$ as, respectively, 2 and -5, which are coincident with those in Figs. 3.3 and 3.5.

Example 2: Fixed-time consensus tracking. Consider a multi-agent system consists of $N = 7$ followers and one leader. The input of the leader is $u_0 = 2\sin(8\pi t)$, so $u_0^{\max} = 2$ can be selected. A connected undirected graph shown in Fig. 3.6 is used as the interaction topology, and its associated matrix $\mathscr{L}$ and matrix $\mathscr{M}$ are denoted as

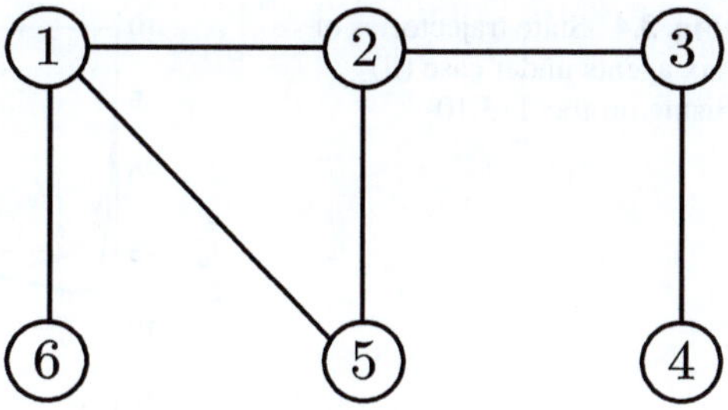

Fig. 3.1 A connected graph

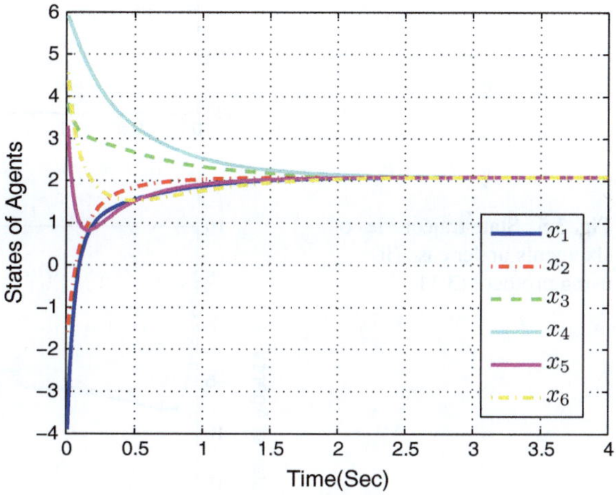

Fig. 3.2 State trajectories of six agents under case (i) using protocol (3.10)

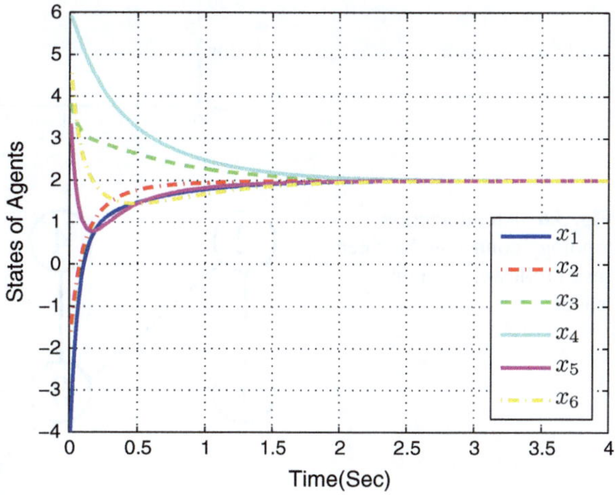

Fig. 3.3 State trajectories of six agents under case (i) using protocol (3.11)

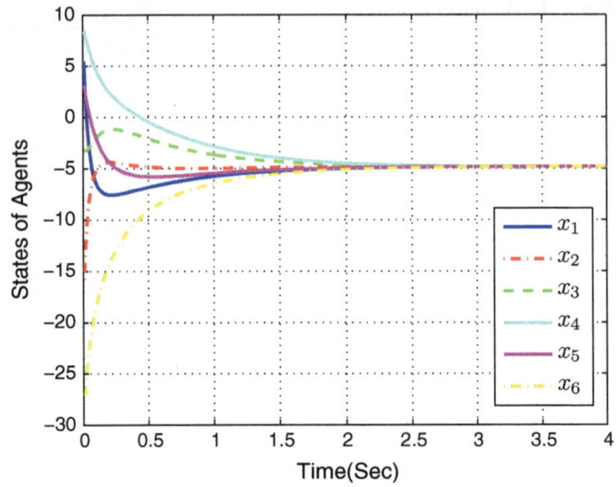

Fig. 3.4 State trajectories of six agents under case (ii) using protocol (3.10)

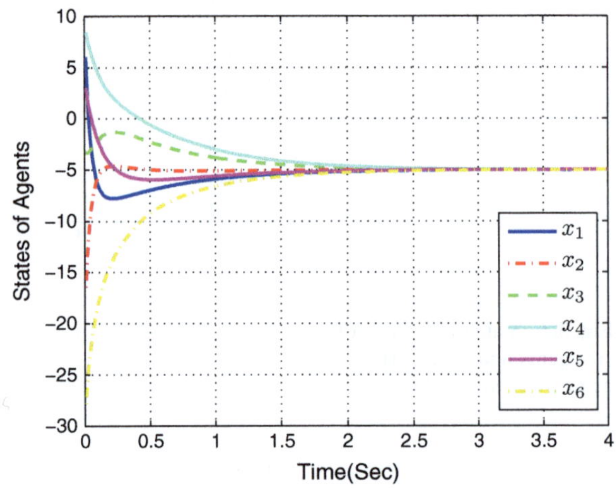

Fig. 3.5 State trajectories of six agents under case (ii) using protocol (3.11)

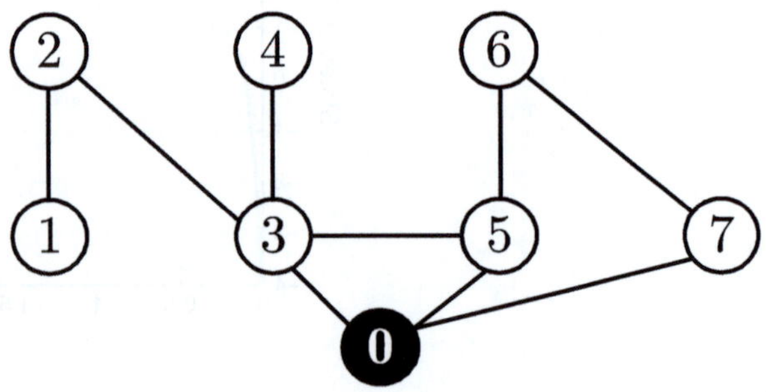

Fig. 3.6 An interaction topology connected between seven followers and one leader

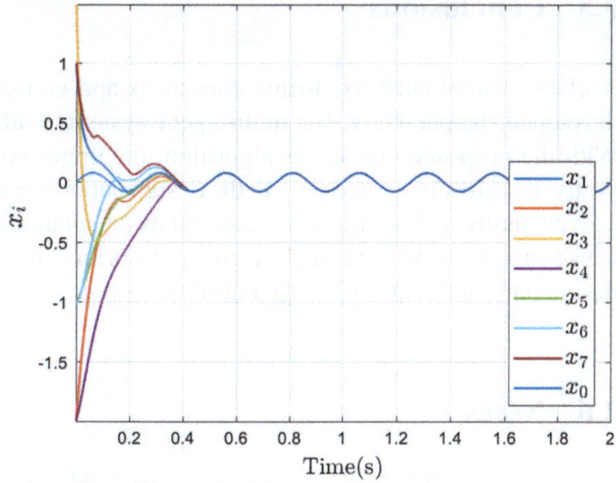

Fig. 3.7 Fixed-time consensus tracking under protocol (3.22)

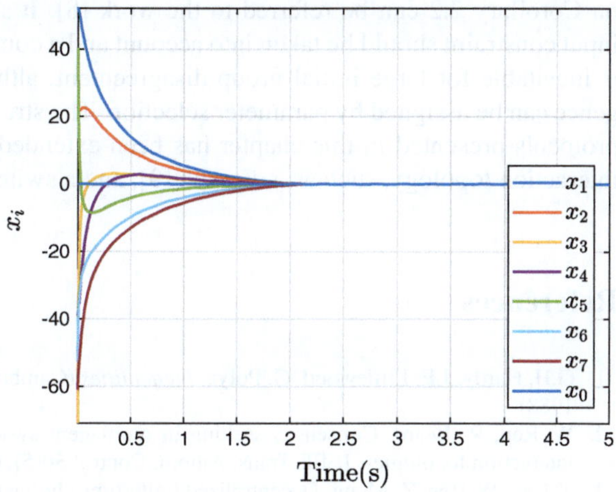

Fig. 3.8 Fixed-time consensus tracking under protocol (3.22)

$$\mathscr{L} = \begin{bmatrix} 0\,1\,0\,0\,0\,0\,0 \\ 1\,0\,1\,0\,0\,0\,0 \\ 0\,1\,0\,1\,1\,0\,0 \\ 0\,0\,1\,0\,0\,0\,0 \\ 0\,0\,1\,0\,0\,1\,0 \\ 0\,0\,0\,0\,1\,0\,1 \\ 0\,0\,0\,0\,0\,1\,0 \end{bmatrix}; \quad \mathscr{M} = \begin{bmatrix} 2 & -1 & 0 & 0 & 0 & 0 & 0 \\ -1 & 2 & -1 & 0 & 0 & 0 & 0 \\ 0 & -1 & 4 & -1 & -1 & 0 & 0 \\ 0 & 0 & -1 & 1 & 0 & 0 & 0 \\ 0 & 0 & -1 & 0 & 3 & -1 & 0 \\ 0 & 0 & 0 & 0 & -1 & 2 & -1 \\ 0 & 0 & 0 & 0 & 0 & -1 & 2 \end{bmatrix}. \tag{3.33}$$

Let $\alpha = 2$, $\beta = 3$, $\gamma = 2$, $p = 5$, $q = 9$ in (3.22). It can be calculated that $T_{\max} = 4.0621$ s. First, set the initial states as $x(0) = [1, -2, 1.5, -2, -1, -1, 1]^T$. The leader evolves from $x_0(0) = 0$. In Fig. 3.7, the settling time is about $t = 0.418$ s under protocol (3.22). Then, change the initial states to $x(0) = [52, 23, -72, -39, 41, -53, -61]^T$. The leader evolves from $x_0(0) = 0$. In Fig. 3.8, the settling time is about $t = 2.052$ s. The settling time for both cases is smaller than $T_{\max}$.

3.5 Conclusions

A class of nonlinear fixed-time consensus and consensus tracking algorithms are developed, respectively, for multi-agent systems with first-order agent dynamics. With the proposed consensus algorithm, the upper bound of settling time for each protocol is closely related only to the design parameters, the algebraic connectivity of a given undirected interaction graph and the agent numbers of a given team. Thus, the consensus reaching time can be estimated a priori off-line via appropriately selecting design parameters for arbitrary initial states.

3.6 Notes

Besides protocols (3.10) and (3.11), another fixed-time consensus protocol based on Corollary 2.2 can be referred to the work [8]. It should be pointed out that an input constraint should be taken into account and a compromise on convergence rate is inevitable for large initial group disagreement, although arbitrarily fast convergence can be assigned by parameter selection. The structure of fixed-time consensus protocols presented in this chapter has been extended to the case of more general interaction topology, such as a digraph [9], and a switching graph [10].

References

1. G.H. Hardy, J.E. Littlewood, G. Polya, *Inequalities* (Cambridge University Press, Cambridge, 1988)
2. W. Ren, R. Beard, Consensus seeking in multiagent systems under dynamically changing interaction topologies. IEEE Trans. Autom. Control **50**(5), 655–661 (2005)
3. Y. Cao, W. Ren, Z. Meng, Decentralized finite-time sliding mode estimators and their applications in decentralized finite-time formation tracking. Syst. Control Lett. **59**(9), 522–529 (2010)
4. H.K. Khalil, *Nonlinear Systems* (Prentice Hall, New Jersey, 2002)
5. R. Olfati-Saber, R.M. Murray, Consensus problems in networks of agents with switching topology and time-delays. IEEE Trans. Autom. Control **49**(9), 1520–1533 (2004)
6. R.A. Horn, C.R. Johnson, *Matrix Analysis*, 2nd edn. (Cambridge University Press, Cambridge, 2012)
7. R. Olfati-Saber, J.A. Fax, R.M. Murray, Consensus and cooperation in networked multi-agent systems. Proc. IEEE **95**(1), 215–233 (2007)
8. Z. Zuo, L. Tie, Distributed robust finite-time nonlinear consensus protocols for multi-agent systems. Int. J. Syst. Sci. **47**(6), 1366–1375 (2016)
9. J. Fu, J. Wang, Finite-time consensus for multi-agent systems with globally bounded convergence time under directed communication graphs. Int. J. Control **90**(9), 1807–1817 (2017)
10. Z. Zuo, W. Yang, L. Tie, D. Meng, Fixed-time consensus for multi-agent systems under directed and switching interaction topology, in *Proceedings of the American Control Conference (ACC)* (Oregon, USA, Portland, 2014), pp. 5133–5138

Chapter 4
Fixed-Time Cooperative Control for Second-Order Multi-Agent Systems

This chapter is concerned with fixed-time consensus tracking for second-order multi-agent systems in networks with directed topology. Nonlinear consensus protocols are constructed with the aid of a sliding surface for each double-integrator agent dynamics. In particular, the proposed method eliminates the singularity and the settling time is assignable for any initial conditions. As a result, consensus protocols are well defined and implementable. More importantly, the convergence time can be estimated or prescribed off-line.

4.1 Problem Formulation

Consider a group of N continuous-time agents as

$$\dot{x}_i(t) = v_i(t),\ \ \dot{v}_i(t) = u_i(t), \tag{4.1}$$

where $x_i \in \mathbb{R}$, $v_i \in \mathbb{R}$, and $u_i \in \mathbb{R}$ are the position, the velocity, and the input (*protocol*) of the ith agent, respectively, and $i \in \Gamma \triangleq \{1, 2, \ldots, N\}$.

Let the consensus reference state evolution be

$$\dot{x}_0(t) = v_0(t),\ \ \dot{v}_0(t) = u_0(t), \tag{4.2}$$

where $x_0 \in \mathbb{R}$, $v_0 \in \mathbb{R}$, and $u_0 \in \mathbb{R}$ are the position, the velocity, and the input of the reference system, respectively. We introduce a virtual leader labeled by 0 with the state $\xi_0 = [x_0, v_0]^T$ for system (4.1). Note, however, that ξ_0 is available not to all agents but to only a portion of agents. Likewise, we define a nonnegative diagonal matrix $\mathscr{B} = \text{diag}(b_1, b_2, \ldots, b_N)$ to indicate the accessibility of ξ_0 by agents, where $b_i = 1$ if ξ_0 is accessible by the ith agent, and $b_i = 0$ otherwise. The directed graph incorporating the leader into $\mathscr{G}$ is denoted by $\bar{\mathscr{G}}$.

Z. Zuo et al., *Fixed-Time Cooperative Control of Multi-Agent Systems*,
https://doi.org/10.1007/978-3-030-20279-8_4

Assumption 4.1.1 $\bar{\mathscr{G}}$ has a spanning tree with the leader being its root vertex, i.e., $\mathscr{B} \neq 0$. □

Assumption 4.1.2 The input $u_0(t)$ of the leader is unavailable to the group members, but its upper-bound $u_0^{\max}$ is accessible by its neighboring agents. □

With a given protocol u_i, the closed-loop system of system (4.1) is said to reach or achieve *fixed-time consensus tracking*, if for $\forall \xi_i(0)$ and $\forall i, j \in \Gamma$, there exists a constant $T_{\max} > 0$ such that the *settling time* $T < T_{\max}$ and

$$\begin{cases} \lim_{t \to T} |\xi_i(t) - \xi_0(t)| \to 0; \\ \xi_i(t) = \xi_0(t), \quad \forall t \geq T, \end{cases} \tag{4.3}$$

where $\xi_i \triangleq [x_i, v_i]^T$, $|\cdot|$ is defined componentwise.

4.2 Nonsingular Fixed-Time Stabilization Control

To clarify the core idea, we first consider fixed-time stabilization control of a single system defined by

$$\dot{z}_1 = z_2, \ \dot{z}_2 = u, \tag{4.4}$$

where $z = [z_1, z_2]^T \in \mathbb{R}^2$ is the system state vector and $u \in \mathbb{R}$ is the control input.

Let m_k, n_k, p_k, q_k be positive odd integers satisfying $m_k > n_k$, $p_1 < q_1 < 2p_1$, $p_2 < q_2$, and $m_1/n_1 - p_1/q_1 > 1$, and α_k, β_k be positive constants, where $k = 1, 2$. Define a C^1 function $\mu_\tau(\cdot) : [0, +\infty) \to [0, 1]$ as

$$\mu_\tau(x) = \begin{cases} \sin\left(\frac{\pi}{2} \cdot \frac{x}{\tau}\right) & \text{if } x \leq \tau; \\ 1 & \text{otherwise}, \end{cases} \tag{4.5}$$

where τ is a positive constant.

To circumvent a singularity issue [1], a sliding surface is proposed:

$$s = z_1 + [\kappa(z_1) \cdot z_2]^{\frac{q_1}{p_1}}, \tag{4.6}$$

where $\kappa(\cdot) : \mathbb{R} \to \mathbb{R}^+$ denotes a scalar positive function given by

$$\kappa(x) = \frac{1}{\alpha_1 x^{m_1/n_1 - p_1/q_1} + \beta_1}. \tag{4.7}$$

In the sequel, the parameter in $\kappa(\cdot)$ will be omitted for simplicity after its first definition. It can be verified that $s = 0$ implies that

$$z_2 = -\alpha_1 z_1^{\frac{m_1}{n_1}} - \beta_1 z_1^{\frac{p_1}{q_1}}.$$

A fixed-time nonlinear control law is defined as

$$
\begin{aligned}
u =& -\frac{1}{\kappa}\left[\alpha_1\left(\frac{m_1}{n_1}-\frac{p_1}{q_1}\right)z_1^{\frac{m_1}{n_1}-\frac{p_1}{q_1}-1}(\kappa z_2)^2-\frac{p_1}{q_1}\kappa^{1-\frac{q_1}{p_1}}z_2^{2-\frac{q_1}{p_1}}\right] \\
& -\frac{p_1}{q_1}\kappa^{-\frac{q_1}{p_1}}\cdot\mu_\tau(z_2^{\frac{q_1}{p_1}-1})\cdot z_2^{1-\frac{q_1}{p_1}}\left(\alpha_2 s^{\frac{m_2}{n_2}}+\beta_2 s^{\frac{p_2}{q_2}}\right), \qquad (4.8)
\end{aligned}
$$

where $\mu_\tau(\cdot)$ is defined in (4.5) with $x = z_2^{q_1/p_1-1}$. The fact that $\sin \omega x \leq x, \forall \omega > 0$ ensures that the input (4.8) is always well defined.

Theorem 4.1 *Consider the second-order system (4.4) with the control feedback law designed as (4.8). Then the origin of the corresponding closed-loop system is fixed-time stable and the settling time estimate is given by*

$$T < T_{\max} := T_1 + T_2 + \varepsilon(\tau), \qquad (4.9)$$

where $T_k \triangleq n_k/[\alpha_k(m_k - n_k)] + q_k/[\beta_k(q_k - p_k)]$, $(k = 1, 2)$, and $\varepsilon(\tau)$ denotes a small time margin related to τ.

Proof Let $f(x(t)) = d(x(t), t) = 0$ and $g(x(t)) = 1$ in (2.20) in Theorem 4.1. The proof is ready to be obtained. □

Remark 4.1 Note that the traveling time $\varepsilon(\tau)$ across $\mathscr{S}_2$ cannot be estimated precisely. However, for sufficiently small τ and $z_1 \neq 0$, integrating both sides of $\dot{z}_2 = u$ subject to $u = -\frac{p_1}{q_1}\kappa^{-p_1/q_1}(\alpha_2 s^{m_2/n_2} + \beta_2 s^{p_2/q_2})$ with $s \to z_1$ yields

$$\varepsilon(\tau) = \frac{q_1}{p_1}[\kappa(|z_1|)]^{\frac{p_1}{q_1}}\frac{1}{\alpha_2|z_1|^{m_2/n_2}+\beta_2|z_1|^{p_2/q_2}}\cdot 2\tau^{\frac{p_1}{q_1-p_1}},$$

where $p_1/(q_1 - p_1) > 1$. Therefore, $\varepsilon(\tau)$ can be made very small by choosing sufficiently small τ. For very small z_1 including the case of $z_1 = 0$, $z_2 \to 0$ in $\mathscr{S}_2$ implies that the system trajectory gets very close to the sliding surface $s = 0$ where the fixed-time convergence is guaranteed, and thus $\varepsilon(\tau) \to 0$. It makes sense to neglect $\varepsilon(\tau)$ in practice for small τ due to the conservative estimate of T_2.

4.3 Nonsingular Fixed-Time Consensus Tracking

Let $e^p = [e_1^p, e_2^p, \ldots, e_N^p]^T$ and $e^v = [e_1^v, e_2^v, \ldots, e_N^v]^T$ be the position and the velocity disagreement vectors, respectively, with their elements defined by $e_i^p = \sum_{j\in\mathscr{N}_i} a_{ij}(x_i - x_j) + b_i(x_i - x_0)$ and $e_i^v = \sum_{j\in\mathscr{N}_i} a_{ij}(v_i - v_j) + b_i(v_i - v_0)$. Taking time derivative yields

$$\dot{e}^p = e^v,\ \dot{e}^v = \mathscr{M}u - \mathscr{B}u_0\mathbf{1}_N, \qquad (4.10)$$

where $\mathscr{M} = \mathscr{L} + \mathscr{B}$ and $u = [u_1, u_2, \ldots, u_N]^T$.

Before proceeding the main result, we need the following Lemma [2] for the proof of convenience in the sequel.

Lemma 4.1 *Consider the system (4.1) and a virtual leader (4.2) under Assumption 4.1.1. If $e^p = 0$ and $e^v = 0$, then $\xi_i = \xi_0$ for all $i \in \Gamma$, i.e.,*

$$[x_1, x_2, \ldots, x_N]^T = \boldsymbol{1}_N x_0; \tag{4.11}$$

$$[v_1, v_2, \ldots, v_N]^T = \boldsymbol{1}_N v_0. \tag{4.12}$$

To solve the fixed-time consensus tracking problem, a nonsingular distributed protocol is derived as

$$u_i = \left(\sum_{j \in \mathscr{N}_i} a_{ij} + b_i\right)^{-1} \left[u_i^d + \sum_{j \in \mathscr{N}_i} a_{ij} u_j - b_i u_0^{\max} \operatorname{sign}(s_i)\right], \tag{4.13}$$

where s_i denotes a sliding surface, defined by

$$s_i = e_i^p + [\kappa_i(e_i^p) \cdot e_i^v]^{\frac{q_1}{p_1}}, \tag{4.14}$$

and u_i^d denotes the virtual control signal, defined by

$$\begin{aligned} u_i^d = & \frac{1}{\kappa_i}\left[\alpha_1\left(\frac{m_1}{n_1} - \frac{p_1}{q_1}\right)(e_i^p)^{\frac{m_1}{n_1} - \frac{p_1}{q_1} - 1}(\kappa_i e_i^v)^2 - \frac{p_1}{q_1}\kappa_i^{1-\frac{q_1}{p_1}}(e_i^v)^{2-\frac{q_1}{p_1}}\right] \\ & - \frac{p_1}{q_1}\kappa_i^{-\frac{q_1}{p_1}}\mu_i^\tau((e_i^v)^{\frac{q_1}{p_1}-1})(e_i^v)^{1-\frac{q_1}{p_1}}\left(\alpha_2 s_i^{\frac{m_2}{n_2}} + \beta_2 s_i^{\frac{p_2}{q_2}}\right), \end{aligned} \tag{4.15}$$

where $\mu_i^\tau(\cdot)$ is defined in (4.5) with $x = (e_i^v)^{q_1/p_1 - 1}$. Note that the threshold parameter τ can be different for each μ_i^τ. Without loss of generality, we choose the same τ for simplicity in this chapter.

Theorem 4.2 *Consider the multi-agent system described by (4.1) and (4.2) in networks with interaction topology satisfying Assumptions 4.1.1 and 4.1.2. The nonsingular distributed protocol proposed in (4.13) achieves fixed-time consensus tracking with the settling time T uniformly bounded by*

$$T < T_{\max} := T_1 + T_2 + \varepsilon(\tau), \tag{4.16}$$

where $T_k = [N^{(m_k - n_k)/2n_k} n_k]/[\alpha_k(m_k - n_k)] + q_k/[\beta_k(q_k - p_k)]$, $(k = 1, 2)$, and $\varepsilon(\tau)$ denotes a small time margin.

Proof (i) Substituting (4.13) into (4.10) yields

$$\dot{e}_i^v = u_i^d - b_i u_0^{\max} \operatorname{sign}(s_i) - b_i u_0. \tag{4.17}$$

Differentiating (4.14) against time and using (4.17), one gets

$$\begin{aligned}\dot{s}_i =& -\mu_i^{\mathcal{T}}((e_i^v)^{\frac{q_1}{p_1}-1})\cdot\left(\alpha_2 s_i^{\frac{m_2}{n_2}}+\beta_2 s_i^{\frac{p_2}{q_2}}\right)\\ &-\frac{q_1}{p_1}\kappa_i^{\frac{q_1}{p_1}}(e_i^v)^{\frac{q_1}{p_1}-1}\left[b_i u_0^{\max}\text{sign}(s_i)+b_i u_0\right].\end{aligned}\tag{4.18}$$

Consider the Lyapunov function $V_2=(1/2)\sum_{i=1}^N s_i^2$ and its time derivative is given by

$$\begin{aligned}\dot{V}_2 =& -\sum_{i=1}^N \mu_i^{\mathcal{T}}((e_i^v)^{\frac{q_1}{p_1}-1})\left(\alpha_2 s_i^{\frac{m_2}{n_2}+1}+\beta_2 s_i^{\frac{p_2}{q_2}+1}\right)\\ &-\kappa_i^{\frac{q_1}{p_1}}(e_i^v)^{\frac{q_1}{p_1}-1}(b_i u_0^{\max}|s_i|+b_i u_0 s_i)\\ \le& -\sum_{i=1}^N \mu_i^{\mathcal{T}}((e_i^v)^{\frac{q_1}{p_1}-1})\left(\alpha_2(s_i^2)^{\frac{m_2+n_2}{2n_2}}+\beta_2(s_i^2)^{\frac{p_2+q_2}{2q_2}}\right)\\ &-\kappa_i^{\frac{q_1}{p_1}}(e_i^v)^{\frac{q_1}{p_1}-1}b_i(u_0^{\max}-|u_0|)|s_i|\\ \le& -\mu_{\mathrm{m}}^{\mathcal{T}}\sum_{i=1}^N\left(\alpha_2(s_i^2)^{\frac{m_2+n_2}{2n_2}}+\beta_2(s_i^2)^{\frac{p_2+q_2}{2q_2}}\right)\\ \le& -\mu_{\mathrm{m}}^{\mathcal{T}}\left[\alpha_2 N^{\frac{n_2-m_2}{2n_2}}(2V_2)^{\frac{m_2+n_2}{2n_2}}+\beta_2(2V_2)^{\frac{p_2+q_2}{2q_2}}\right],\end{aligned}$$

where $\kappa_i^{q_1/p_1}\cdot(e_i^v)^{q_1/p_1-1}\ge 0$, $\mu_{\mathrm{m}}^{\mathcal{T}}\triangleq\min\{\mu_1^{\mathcal{T}},\ldots,\mu_N^{\mathcal{T}}\}$. Note that Lemmas 3.1 and 3.2 are used in the last inequality. If $V_2\ne 0$, let $y_2=\sqrt{2V_2}$. Then $\dot{y}_2=\dot{V}_2/\sqrt{2V_2}$. We have that

$$\dot{y}_2(t)\le -\mu_{\mathrm{m}}^{\mathcal{T}}\left(\alpha_2 N^{\frac{n_2-m_2}{2n_2}}y_2^{\frac{m_2}{n_2}}(t)+\beta_2 y_2^{\frac{p_2}{q_2}}(t)\right),$$

Since $\mu_{\mathrm{m}}^{\mathcal{T}}=1$ if $\min\{|e_1^v|,|e_2^v|,\ldots,|e_N^v|\}>\tau^{p_1/(q_1-p_1)}$, by using Theorem 4.1 and *Comparison Principle* of differential equations, we obtain that $\lim_{t\to t_2}V_2(s_i)=0$, which implies that $\lim_{t\to t_2}s_i=0$ for all $i\in\Gamma$, where the settling time t_2 is bounded by $t_2<T_2+\varepsilon(\tau)$.

(ii) Note that on the sliding surface $s_i=0$, Eq. (4.14) can be written as

$$\dot{e}_i^p=-\alpha_1(e_i^p)^{\frac{m_1}{n_1}}-\beta_1(e_i^p)^{\frac{p_1}{q_1}}\tag{4.19}$$

for all $i\in\Gamma$. Consider the Lyapunov function $V_1=(1/2)\sum_{i=1}^N(e_i^p)^2$ and its time derivative is

$$\begin{aligned}\dot{V}_1 =& -\alpha_1\sum_{i=1}^N(e_i^p)^{\frac{m_1+n_1}{n_1}}-\beta_1\sum_{i=1}^N(e_i^p)^{\frac{p_1+q_1}{q_1}}\\ \le& -\alpha_1 N^{\frac{n_1-m_1}{2n_1}}(2V_1)^{\frac{m_1+n_1}{2n_1}}-\beta_1(2V_1)^{\frac{p_1+q_1}{2q_1}},\end{aligned}$$

where Lemmas 3.1 and 3.2 are applied. If $V_1 \neq 0$, let $y_1 = \sqrt{2V_1}$. Then we have that

$$\dot{y}_1(t) \leq -\alpha_1 N^{\frac{n_1-m_1}{2n_1}} y_1^{\frac{m_1}{n_1}}(t) - \beta_1 y_1^{\frac{p_1}{q_1}}(t).$$

Similarly as (i), one obtains that $\lim_{t\to t_1} e_i^p = 0$ with the settling time t_1 bounded by $t_1 < T_1$ for all $i \in \Gamma$. Thus, e^p converges to zero within the period $T < T_1 + T_2 + \varepsilon(\tau)$. Applying Lemma 4.1, one can conclude that fixed-time consensus tracking is achieved. □

To incorporate the interaction topology into the sliding surface, we modify the sliding variables as

$$\bar{s}_i = \bar{e}_i^p + [\kappa_i(\bar{e}_i^p) \cdot e_i^v]^{\frac{q_1}{p_1}}, \tag{4.20}$$

where $\bar{e}_i^p = \sum_{j\in\mathcal{N}_i} \hat{a}_{ij}(e_i^p - e_j^p) + b_i e_i^p$ with $\hat{a}_{ij} = (a_{ij} + a_{ji})/2$ for all $i, j \in \Gamma$. Let $\overline{\mathcal{L}} = (\mathcal{L} + \mathcal{L}^T)/2$ be the Laplacian matrix of $\mathcal{G}$ with $\overline{\mathcal{A}} \triangleq [\hat{a}_{ij}]_{n\times n}$. Thus, the corresponding protocol for fixed-time consensus tracking is obtained as

$$u_i = \Big(\sum_{j\in\mathcal{N}_i} a_{ij} + b_i\Big)^{-1}\Big[\bar{u}_i^d + \sum_{j\in\mathcal{N}_i} a_{ij}u_j - b_i u_0^{\max}\mathrm{sign}(\bar{s}_i)\Big], \tag{4.21}$$

where the virtual signal $\bar{u}_i^d$ is given by

$$\begin{aligned}\bar{u}_i^d = &\frac{1}{\kappa_i}\left[\alpha_1\Big(\frac{m_1}{n_1} - \frac{p_1}{q_1}\Big)(\bar{e}_i^p)^{\frac{m_1}{n_1}-\frac{p_1}{q_1}-1}(\kappa_i e_i^v)^2 - \frac{p_1}{q_1}\kappa_i^{1-\frac{q_1}{p_1}}(e_i^v)^{2-\frac{q_1}{p_1}}\right] \\ &- \frac{p_1}{q_1}\kappa_i^{-\frac{q_1}{p_1}}\mu_i^T((e_i^v)^{\frac{q_1}{p_1}-1})(e_i^v)^{1-\frac{q_1}{p_1}}\Big(\alpha_2\bar{s}_i^{\frac{m_2}{n_2}} + \beta_2\bar{s}_i^{\frac{p_2}{q_2}}\Big).\end{aligned} \tag{4.22}$$

Theorem 4.3 *Consider the multi-agent system described by (4.1) and (4.2) in networks with interaction topology satisfying Assumptions 4.1.1 and 4.1.2. The nonsingular distributed protocol proposed in (4.21) achieves fixed-time consensus tracking with the settling time bounded by*

$$T < T_{\max} := T_1 + T_2 + \varepsilon(\tau), \tag{4.23}$$

where

$$T_1 = \frac{N^{(m_1-n_1)/2n_1}}{\underline{\lambda}^{(m_1+n_1)/2n_1}\alpha_1}\frac{n_1}{m_1-n_1} + \frac{1}{\underline{\lambda}^{(p_1+q_1)/2q_1}\beta_1}\frac{q_1}{q_1-p_1};$$

$$T_2 = \frac{N^{(m_2-n_2)/2n_2}}{\alpha_2}\frac{n_2}{m_2-n_2} + \frac{1}{\beta_2}\frac{q_2}{q_2-p_2}$$

with $\underline{\lambda}$ denoting the smallest eigenvalue of matrix $\overline{\mathcal{L}} + \mathcal{B}$, and $\varepsilon(\tau)$ denoting a small time margin.

Proof The proof follows the same line of Theorem 4.2.

(i) Similar to (i) of the proof in Theorem 4.2, we have that $\lim_{t\to t_2} s_i = 0$ for all $i \in \Gamma$, where $t_2 < T_2 + \varepsilon(\tau)$.

(ii) $s_i = 0$ in (4.20) implies that $e^v = -\alpha_1(\bar{e}^p)^{m_1/n_1} - \beta_1(\bar{e}^p)^{p_1/q_1}$, where $(\cdot)^{m_1/n_1}$ and $(\cdot)^{p_1/q_1}$ are defined componentwise. Let $\bar{e}^v \triangleq [\bar{e}^v_1, \bar{e}^v_2, \ldots, \bar{e}^v_N]^T$. We have

$$\bar{e}^v = \dot{\bar{e}}^p = -\left(\overline{\mathscr{L}} + \mathscr{B}\right)[\alpha_1(\bar{e}^p)^{m_1/n_1} + \beta_1(\bar{e}^p)^{p_1/q_1}]$$

Consider the Lyapunov function $V_1 = (\bar{e}^p)^T(\overline{\mathscr{L}} + \mathscr{B})^{-1}(\bar{e}^p)/2$ and its time derivative is derived as

$$\begin{aligned}
\dot{V}_1 &= -(\bar{e}^p)^T\left[\alpha_1(\bar{e}^p)^{\frac{m_1}{n_1}} + \beta_1(\bar{e}^p)^{\frac{p_1}{q_1}}\right]\\
&= -\sum_{i=1}^{N}\left[\alpha_1(\bar{e}^p_i)^{\frac{m_1}{n_1}+1} + \beta_1(\bar{e}^p_i)^{\frac{p_1}{q_1}+1}\right]\\
&\le -\left[\alpha_1 N^{\frac{n_1-m_1}{2n_1}}\left(\sum_{i=1}^{N}(\bar{e}^p_i)^2\right)^{\frac{m_1+n_1}{2n_1}} + \beta_1\left(\sum_{i=1}^{N}(\bar{e}^p_i)^2\right)^{\frac{p_1+q_1}{2q_1}}\right]\\
&\le -\left[\alpha_1 N^{\frac{n_1-m_1}{2n_1}}(2\underline{\lambda}V_1)^{\frac{m_1+n_1}{2n_1}} + \beta_1(2\underline{\lambda}V_1)^{\frac{p_1+q_1}{2q_1}}\right].
\end{aligned}$$

Let $y_1 = \sqrt{2V_1}$. It follows that

$$\dot{y}_1(t) \le -\alpha_1 N^{\frac{n_1-m_1}{2n_1}}\underline{\lambda}^{\frac{m_1+n_1}{2n_1}} y_1^{\frac{m_1}{n_1}}(t) - \beta_1\underline{\lambda}^{\frac{p_1+q_1}{2q_1}} y_1^{\frac{p_1}{q_1}}(t).$$

Similar to (ii) of the proof in Theorem 4.2, we have that $\lim_{t\to t_1}\bar{e}^p = 0$, and the settling time t_1 is bounded by

$$t_1 < \frac{N^{(m_1-n_1)/2n_1}}{\underline{\lambda}^{m_1+n_1/2n_1}\alpha_1}\frac{n_1}{m_1-n_1} + \frac{1}{\underline{\lambda}^{p_1+q_1/2q_1}\beta_1}\frac{q_1}{q_1-p_1} \triangleq T_1.$$

Note that $\bar{e}^p = (\overline{\mathscr{L}} + \mathscr{B})e^p$ and $\overline{\mathscr{L}} + \mathscr{B} > 0$ by Lemma 2.5, which implies that $\lim_{t\to T} e^p = 0$ and $T < T_1 + T_2 + \varepsilon(\tau)$. Thus, (4.11) and (4.12) follow. This competes the proof. □

4.4 A Numerical Example

Consider a group of six agents in a network with the interaction graph shown in Fig. 4.1. The time-varying control input of the virtual leader (4.2) is designed as $u_0 = -\sin(x_0)/(1 + \exp(-t))$ verifying $u_0^{\max} = 1$. Consider the initial scenario that

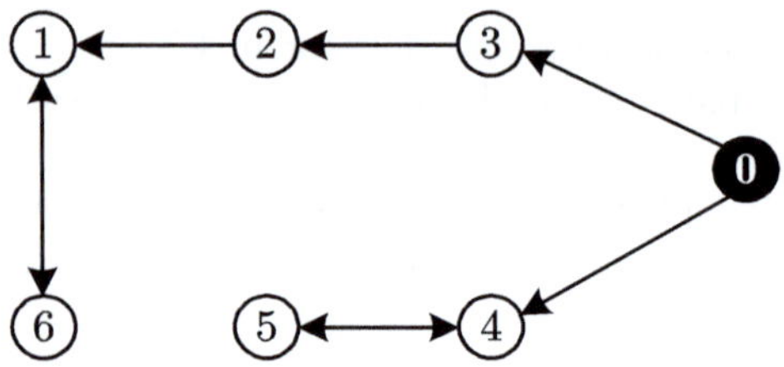

Fig. 4.1 The interaction graph

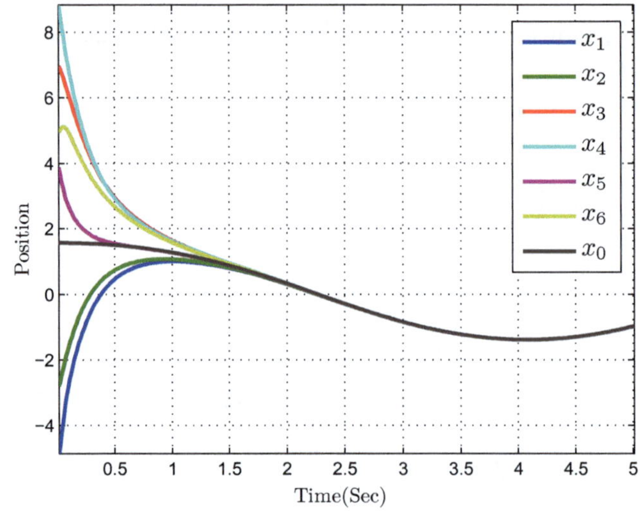

(a) Protocol (4.13)

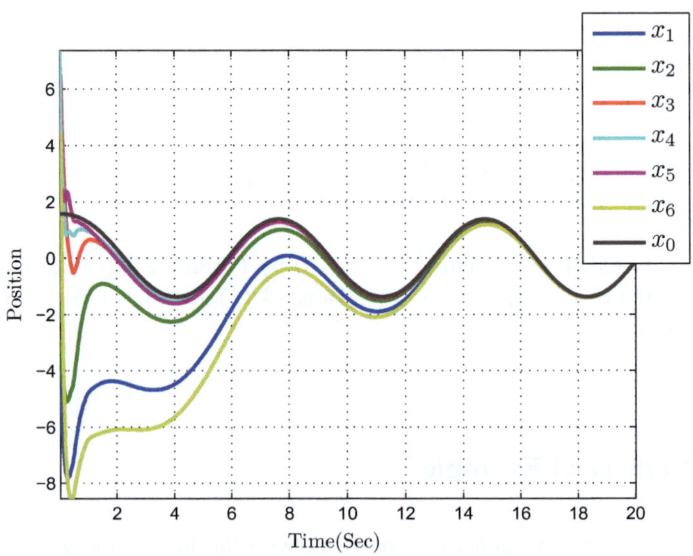

(b) Protocol (4.21)

Fig. 4.2 Consensus tracking results

$x(0) = [-5, -3, 7, 9, 4, 5]^T$ and $v(0) = 0$ are set for the group of agents (4.1) and $x_0(0) = \pi/2$ and $v_0(0) = 0$ for the virtual leader (4.2). The design parameters $\alpha_1 = \beta_1 = \alpha_2 = \beta_2 = 2, m_1 = 9, n_1 = 5, p_1 = 7, q_1 = 9, m_2 = 11, n_2 = 9, p_2 = 5, q_2 = 7$, and $\tau = 0.1$ are set for protocols (4.13) and (4.21). It can be verified that the parameter constraints that $m_1/n_1 - p_1/q_1 > 1$ and $p_1/q_1 > 1/2$ are satisfied. The bounds for settling time T in (4.16) and (4.23) are 8.025 s and 24.985 s, respectively. Since $\varepsilon_2 = 7/9 < 1$, a less conservative bounds can be calculated as 7.889 s and 24.425 s.

The simulation results in Fig. 4.2 show that the consensus tracking objective is achieved within fixed time. Note that the settling times under the two protocols are about, respectively, 2 s and 20 s, demonstrating the performance claimed in Theorems 4.2 and 4.3. It can be further observed from Fig. 4.2a that each agent in the group reaches the reference state directly, while from Fig. 4.2b the group seeks consensus among neighbors during transition and reaches the reference state together.

4.5 Conclusions

A guaranteed settling time bound independent of initial conditions is obtained and fixed-time consensus is achieved. Nonlinear consensus protocols are constructed with the aid of a sliding surface for each double-integrator agent dynamics. In particular, the proposed framework eliminates the singularity and the settling time is assignable for any initial conditions. A numerical example is given to demonstrate the effectiveness of the proposed protocol.

4.6 Notes

This chapter extends an idea in [3, 4] and develops fixed-time consensus protocols for second-order multi-agent networks with directed information flow. The main materials of this chapter are based on [5]. It is worth mentioning that the settling time estimate (4.23) also relies on the interaction topology property (i.e., $\underline{\lambda}$). Furthermore, it is worth pointing out that introducing topology into the definition of sliding variables, as shown in (4.21), involves the state interaction among neighboring agents during the transition, while the protocol in (4.13) achieves the state agreement of each agent directly to the reference state.

References

1. Y. Feng, X. Yu, Z. Man, On nonsingular terminal sliding-mode control of nonlinear systems. Automatica **49**(6), 1715–1722 (2013)
2. S. Khoo, L. Xie, Z. Man, Robust finite-time consensus tracking algorithm for multirobot systems. IEEE/ASME Trans. Mechatron. **14**(2), 219–228 (2009)
3. Z. Zuo, L. Tie, A new class of finite-time nonlinear consensus protocols for multi-agent systems. Int. J. Control **87**(2), 363–370 (2014)
4. Z. Zuo, L. Tie, Distributed robust finite-time nonlinear consensus protocols for multi-agent systems. Int. J. Syst. Sci. **47**(6), 1366–1375 (2016)
5. Z. Zuo, Nonsingular fixed-time consensus tracking for second-order multi-agent networks. Automatica **54**, 305–309 (2015)

Chapter 5
Fixed-Time Cooperative Control for High-Order Multi-Agent Systems

This chapter addresses a fixed-time leader–follower consensus problem for high-order integrator multi-agent systems subject to matched external disturbances. A cascade control structure, based on a fixed-time distributed observer, is developed to achieve fixed-time consensus tracking control. A numerical example is given to illustrate the effectiveness and performance of the proposed control structure with respect to different initial conditions.

5.1 Problem Formulation

Consider $N+1$ agents with one leader and N followers labeled by 0 and $i = 1, 2, \ldots, N$, respectively. The leader is described by

$$\begin{cases} \dot{x}_{0,1}(t) = x_{0,2}(t), \\ \dot{x}_{0,2}(t) = x_{0,3}(t), \\ \qquad \vdots \\ \dot{x}_{0,n}(t) = u_0(t), \end{cases} \tag{5.1}$$

where $x_0 = [x_{0,1}, x_{0,2}, \ldots, x_{0,n}]^T \in \mathbb{R}^n$ is the state vector, and $u_0 \in \mathbb{R}$ is the control input of the leader. The ith follower agent is described by

$$\begin{cases} \dot{x}_{i,1}(t) = x_{i,2}(t), \\ \dot{x}_{i,2}(t) = x_{i,3}(t), \\ \qquad \vdots \\ \dot{x}_{i,n}(t) = u_i(t) + \Delta_i(t), \end{cases} \tag{5.2}$$

Z. Zuo et al., *Fixed-Time Cooperative Control of Multi-Agent Systems*,
https://doi.org/10.1007/978-3-030-20279-8_5

where $x_i = [x_{i,1}, x_{i,2}, \ldots, x_{i,n}]^T \in \mathbb{R}^n$ and $u_i \in \mathbb{R}$ are the state vector and the control input, respectively, of the ith agent, and Δ_i is the lumped uncertainties in the ith agent. Note that the term Δ_i may result from external disturbances, uncertain dynamics, or uncertainties in earlier states after differentiation.

Several reasonable assumptions are made for consensus tracking design discussed in this chapter.

Assumption 5.1.1 Graph $\mathscr{G}$ is connected and at least one follower in graph $\mathscr{G}$ can get access to the states of the leader, i.e., $\mathscr{B} \neq 0$. □

Assumption 5.1.2 The input u_0 of the leader is unknown to any followers but its upper bound, denoted by $u_0^{\max}$, can be accessible by the ith follower if $b_i \neq 0$, $i \in \Gamma$. □

Assumption 5.1.3 The unknown disturbance $\Delta_i(t)$ in (5.2) is uniformly bounded by a known constant δ, i.e., $|\Delta_i(t)| \leq \delta$ for all $i \in \Gamma$. □

Remark 5.1 Assumption 5.1.1 is necessary for solving a consensus tracking problem, and more discussion on this assumption can be referred to [1].

Remark 5.2 Assumption 5.1.2 is mild for physical systems since an input of the leader is known and its upper bound can be obtained *a prior*. Assumption 5.1.3 is made for robust control design and is conventional in sliding mode control [2].

The control objective is to design a distributed protocol u_i by using only relative information for each follower subject to external disturbances such that fixed-time consensus is achieved, i.e., for $\forall x_i(0)$, $\forall i \in \Gamma$, there exists a constant $T_{\max}$ such that

$$\begin{cases} \lim_{t \to T_{\max}} \|x_i(t) - x_0(t)\|_2 = 0; \\ x_i(t) = x_0(t), \quad \forall t > T_{\max}. \end{cases} \tag{5.3}$$

5.2 Fixed-Time Consensus Tracking

In this section, a fixed-time consensus tracking protocol, based on a cascaded structure, is developed for multi-agent systems with high-order integrator dynamics.

5.2.1 *Distributed Observer*

Consider the information of the leader to be available not to all followers but to only a portion of them. We first propose an observer for each follower to get an estimate of the state of the leader.

Denote by $\hat{x}_{0,k}^i$ the estimate of the leader's state $x_{0,k}$, $k = 1, 2, \ldots, n$ for the ith follower, $i \in \Gamma$. A distributed fixed-time observer takes the following structure:

$$\begin{aligned}\dot{\hat{x}}_{0,k}^{i} &= \hat{x}_{0,k+1}^{i} \\ &\quad + \alpha_k \mathrm{sign}\left(\sum_{j=1}^{N} a_{ij}(\hat{x}_{0,k}^{j} - \hat{x}_{0,k}^{i}) + b_i(x_{0,k} - \hat{x}_{0,k}^{i})\right) \\ &\quad + \beta_k \left[\sum_{j=1}^{N} a_{ij}(\hat{x}_{0,k}^{j} - \hat{x}_{0,k}^{i}) + b_i(x_{0,k} - \hat{x}_{0,k}^{i})\right]^2, \\ &\qquad (k = 1, 2, \ldots, n-1), \\ \dot{\hat{x}}_{0,n}^{i} &= \alpha_n \mathrm{sign}\left(\sum_{j=1}^{N} a_{ij}(\hat{x}_{0,n}^{j} - \hat{x}_{0,n}^{i}) + b_i(x_{0,n} - \hat{x}_{0,n}^{i})\right) \\ &\quad + \beta_n \left[\sum_{j=1}^{N} a_{ij}(\hat{x}_{0,n}^{j} - \hat{x}_{0,n}^{i}) + b_i(x_{0,n} - \hat{x}_{0,n}^{i})\right]^2. \end{aligned} \tag{5.4}$$

Let the observation errors be

$$\tilde{x}_{0,k}^{i} = \hat{x}_{0,k}^{i} - x_{0,k}. \tag{5.5}$$

With (5.5), the observation error system can be derived as

$$\begin{aligned}\dot{\tilde{x}}_{0,k}^{i} &= \tilde{x}_{0,k+1}^{i} + \alpha_k \mathrm{sign}\left(\sum_{j=1}^{N} a_{ij}(\tilde{x}_{0,k}^{j} - \tilde{x}_{0,k}^{i}) - b_i \tilde{x}_{0,k}^{i}\right) \\ &\quad + \beta_k \left[\sum_{j=1}^{N} a_{ij}(\tilde{x}_{0,k}^{j} - \tilde{x}_{0,k}^{i}) - b_i \tilde{x}_{0,k}^{i}\right]^2, \\ &\qquad (k = 1, 2, \ldots, n-1), \\ \dot{\tilde{x}}_{0,n}^{i} &= \alpha_n \mathrm{sign}\left(\sum_{j=1}^{N} a_{ij}(\tilde{x}_{0,n}^{j} - \tilde{x}_{0,n}^{i}) - b_i \tilde{x}_{0,n}^{i}\right) \\ &\quad + \beta_n \left[\sum_{j=1}^{N} a_{ij}(\tilde{x}_{0,n}^{j} - \tilde{x}_{0,n}^{i}) - b_i \tilde{x}_{0,n}^{i}\right]^2 - u_0. \end{aligned} \tag{5.6}$$

Let $z_i = [\tilde{x}_{0,i}^{1}, \tilde{x}_{0,i}^{2}, \ldots, \tilde{x}_{0,i}^{N}]^T$ for $i = 1, 2, \ldots, n$. From (5.6), a compact form can be written as

$$\begin{aligned}\dot{z}_k &= z_{k+1} - \alpha_k \mathrm{sign}\,(\mathcal{M} z_k) - \beta_k \lceil \mathcal{M} z_k \rfloor^2, \\ &\qquad (k = 1, 2, \ldots, n-1), \\ \dot{z}_n &= -\alpha_n \mathrm{sign}\,(\mathcal{M} z_n) - \beta_n \lceil \mathcal{M} z_n \rfloor^2 - \mathbf{1} u_0. \end{aligned} \tag{5.7}$$

The fixed-time convergence property of the estimate errors is summarized in the following theorem.

Theorem 5.1 *Under Assumptions 5.1.1 and 5.1.2, if the observer gains satisfy*

$$\beta_k = \frac{\varepsilon\sqrt{N}}{(2\lambda_{\min}(\mathcal{M}))^{\frac{3}{2}}}, \quad \forall k = 1, 2, \ldots, n, \tag{5.8}$$

$$\alpha_k = \varepsilon\sqrt{\frac{\lambda_{\max}(\mathcal{M})}{2\lambda_{\min}(\mathcal{M})}}, \quad \forall k = 1, 2, \ldots, n-1, \tag{5.9}$$

$$\alpha_n = u_0^{\max} + \varepsilon\sqrt{\frac{\lambda_{\max}(\mathcal{M})}{2\lambda_{\min}(\mathcal{M})}}, \tag{5.10}$$

where $\varepsilon > 0$*, then the distributed observer* (5.4) *achieves the convergence of the observation errors to zero in a fixed time which is bounded by*

$$T_o := \frac{n\pi}{\varepsilon}. \tag{5.11}$$

Proof The proof consists of two steps. First, we show that the observation errors will not escape to infinity in finite time, i.e., the observation errors are bounded on any time interval $[0, t]$. Then fixed-time stability of (5.7) will be proved in a recursive manner.

(i) Construct the following Lyapunov function:

$$V = \sum_{i=1}^{n} V_i = \frac{1}{2}\sum_{i=1}^{n} z_i^T \mathcal{M} z_i, \tag{5.12}$$

where $V_i = \frac{1}{2} z_i^T \mathcal{M} z_i, i = 1, 2, \ldots, n$, which are positive definite under Assumption 5.1.1 by Lemma 2.5. Differentiate V along (5.7), and one obtains

$$\begin{aligned}\dot{V} &= \sum_{i=1}^{n-1} z_i^T \mathcal{M} z_{i+1} - z_n^T \mathcal{M} \mathbf{1} u_0 - \sum_{i=1}^{n} \alpha_i z_i^T \mathcal{M} \mathrm{sign}(\mathcal{M} z_i) - \sum_{i=1}^{n} \beta_i z_i^T \mathcal{M} \lceil \mathcal{M} z_i \rfloor^2 \\ &\le \frac{1}{2}\sum_{i=1}^{n-1} z_i^T \mathcal{M} z_i + \frac{1}{2}\sum_{i=1}^{n-1} z_{i+1}^T \mathcal{M} z_{i+1} - \sum_{i=1}^{n-1} \alpha_i \|\mathcal{M} z_i\|_1 - \sum_{i=1}^{n} \beta_i \|\mathcal{M} z_i\|_2 \\ &\quad - (\alpha_n - u_0^{\max})\|\mathcal{M} z_n\|_1 \\ &\le 2V, \end{aligned} \tag{5.13}$$

where inequality $2a^T b \leq a^T a + b^T b$ with vectors a and b is used to obtain the second inequality. Thus, V is bounded on any time interval $[0, t]$, which implies from (5.12) that all z_i will not escape to infinity in finite time. Moreover, it follows from (5.5) that all states of the observer (5.4) are bounded on any finite-time interval if the state x_0 of the leader in (5.1) will not escape to infinity in finite time.

(ii) Consider the following Lyapunov function:

$$V_n = \frac{1}{2} z_n^T \mathscr{M} z_n. \tag{5.14}$$

Its time derivative results in

$$\begin{aligned}\dot{V}_n &= z_n^T \mathscr{M} \left(-\alpha_n \mathrm{sign}\,(\mathscr{M} z_n) - \beta_n \lceil \mathscr{M} z_n \rfloor^2\right) - z_n^T \mathscr{M} \mathbf{1} u_0 \\ &\leq -(\alpha_n - u_0^{\max}) \|\mathscr{M} z_n\|_1 - \beta_n N^{-\frac{1}{2}} \left(2\lambda_{\min}(\mathscr{M})\right)^{\frac{3}{2}} V_n^{\frac{3}{2}} \\ &\leq -\varepsilon V_n^{\frac{1}{2}} - \varepsilon V_n^{\frac{3}{2}}.\end{aligned} \tag{5.15}$$

Hence, Theorem 2.4 guarantees that z_n is fixed-time stable at the origin with the settling time bounded by $T_1 = \frac{\pi}{\varepsilon}$.

After the convergence of z_n, the dynamics of z_{n-1} reduces to

$$\dot{z}_{n-1} = -\alpha_{n-1} \mathrm{sign}\,(\mathscr{M} z_{n-1}) - \beta_{n-1} \lceil \mathscr{M} z_{n-1} \rfloor^2.$$

Similar, one obtains that z_{n-1} converges to zero in a fixed-time bounded by $T_2 = 2T_1$. Recursively, one gets that z_1 converges to zero within a fixed-time horizon bounded by $T_o := T_n = nT_1$. This completes the proof. □

Remark 5.3 It is worth noting that the classical observer (e.g., the ones in [3–5] for consensus problem) is used to reconstruct the state of a system based on incomplete measurements. Different from the concept of conventional observer, the observer (5.4) is proposed to reconstruct the leader state in a distributed manner for each follower since only a portion of followers in a group can get access to the leader state. Theorem 5.1 guarantees that the observer (5.4) could recover the leader state within a fixed time if Assumption 5.1.1 holds. Thus, the dynamic system (5.4) is still referred to as an "observer" without much confusion. Furthermore, it should be highlighted that if $x_{0,k}$ is transmitted from agent i to agent j, instead of $\hat{x}_{0,k}^i$, the scheme becomes a centralized one. Then issues of security and robustness to agent and communication link failures clearly make this centralized strategy prohibitive. Hence, in this chapter, the online reconstruction of the leader state for all followers is proposed to avoid such issues since only local information is transmitted.

Remark 5.4 The conclusion in Theorem 5.1 implies that $\hat{x}_{0,k}^i = x_{0,k}$ for all $t \geq T_o$, $k = 1, 2, \ldots, n$, $i \in \Gamma$. In other words, each follower is able to accurately estimate the state of the leader after a period of time and thus we can use $\hat{x}_{0,k}^i$ in the protocol design based on a cascaded structure.

Remark 5.5 The settling time estimate (5.11) derived in Theorem 5.1 is independent of initial observation errors. To be specific, for any given desired T_o, ε will be specified by (5.11) accordingly if the order of the multi-agent system is fixed, but the observer parameters (5.8)–(5.10) will be then tuned according to the prescribed ε, the network topology and the number of agents in a group. In other words, (5.11) provides an explicit algorithm for adjusting the observer parameters to predefine the settling time estimate.

5.2.2 *Consensus Tracking*

Define the tracking error

$$e_{i,k} = x_{i,k} - \hat{x}_{0,k}^i = x_{i,k} - x_{0,k} - \tilde{x}_{0,k}^i, \tag{5.16}$$

where $\tilde{x}_{0,k}^i$ is defined in (5.5), $k = 1, 2, \ldots, n$ and $i \in \Gamma$. With (5.6), the time derivative of the tracking error $e_{i,k}$ can be derived as

$$\begin{aligned}
\dot{e}_{i,k} &= e_{i,k+1} - \alpha_k \mathrm{sign}\left(\sum_{j=1}^{N} a_{ij}(\tilde{x}_{0,k}^j - \tilde{x}_{0,k}^i) - b_i \tilde{x}_{0,k}^i\right) \\
&\quad - \beta_k \left[\sum_{j=1}^{N} a_{ij}(\tilde{x}_{0,k}^j - \tilde{x}_{0,k}^i) - b_i \tilde{x}_{0,k}^i\right]^2, \\
&\qquad (k = 1, 2, \ldots, n-1), \\
\dot{e}_{i,n} &= u_i + \Delta_i - \alpha_n \mathrm{sign}\left(\sum_{j=1}^{N} a_{ij}(\tilde{x}_{0,n}^j - \tilde{x}_{0,n}^i) - b_i \tilde{x}_{0,n}^i\right) \\
&\quad - \beta_n \left[\sum_{j=1}^{N} a_{ij}(\tilde{x}_{0,n}^j - \tilde{x}_{0,n}^i) - b_i \tilde{x}_{0,n}^i\right]^2.
\end{aligned} \tag{5.17}$$

By Theorem 5.1, one obtains that $\tilde{x}_{0,k}^i = 0$ for all $t \geq T_o$. Let $e_k = [e_{1,k}, e_{2,k}, \ldots, e_{N,k}]^T$. From (5.1) and (5.2), for all $t \geq T_o$, the tracking error dynamics can be written in a compact form:

$$\begin{cases}
\dot{e}_1(t) = e_2(t), \\
\dot{e}_2(t) = e_3(t), \\
\quad \vdots \\
\dot{e}_n(t) = U(t) + \Delta(t),
\end{cases} \tag{5.18}$$

where $U = [u_1, u_2, \ldots, u_N]^T$, $\Delta = [\Delta_1, \Delta_2, \ldots, \Delta_N]^T$.

To solve the fixed-time consensus tracking problem, we propose the following integral sliding surface for each follower:

$$s_i(t) = e_{i,n}(t) + \sum_{j=1}^{n} k_i \int_0^t \lceil e_{i,j}(\tau) \rfloor^{\gamma_i} \mathrm{d}\tau + \sum_{j=1}^{n} \bar{k}_i \int_0^t \lceil e_{i,j}(\tau) \rfloor^{\bar{\gamma}_i} \mathrm{d}\tau, \tag{5.19}$$

where $i \in \Gamma$, the parameters k_i, $\bar{k}_i$, γ_i, and $\bar{\gamma}_i$ are selected as given in Theorem 2.11. Differentiating s_i against time yields

$$\begin{aligned} \dot{s}_i(t) = u_i(t) &+ \sum_{j=1}^{n} \left(k_i \lceil e_{i,j}(t) \rfloor^{\gamma_i} + \bar{k}_i \lceil e_{i,j}(t) \rfloor^{\bar{\gamma}_i} \right) \\ &+ \Delta_i(t) - \alpha_n \mathrm{sign} \left(\sum_{j=1}^{N} a_{ij} (\tilde{x}_{0,n}^j - \tilde{x}_{0,n}^i) - b_i \tilde{x}_{0,n}^i \right) \\ &- \beta_n \left\lceil \sum_{j=1}^{N} a_{ij} (\tilde{x}_{0,n}^j - \tilde{x}_{0,n}^i) - b_i \tilde{x}_{0,n}^i \right\rfloor^2, \end{aligned} \tag{5.20}$$

and, for all $t \geq T_o$, dynamics $\dot{s}_i$ in (5.20) reduces to

$$\dot{s}_i(t) = u_i(t) + \sum_{j=1}^{n} \left(k_i \lceil e_{i,j}(t) \rfloor^{\gamma_i} + \bar{k}_i \lceil e_{i,j}(t) \rfloor^{\bar{\gamma}_i} \right) + \Delta_i(t). \tag{5.21}$$

In view of (5.21), a fixed-time consensus tracking protocol can be prescribed as

$$u_i = -\sum_{j=1}^{n} \left(k_i \lceil e_{i,j} \rfloor^{\gamma_i} + \bar{k}_i \lceil e_{i,j} \rfloor^{\bar{\gamma}_i} \right) - \rho \mathrm{sign}(s_i) - k_i \left(\lceil s_i \rfloor^{1+\frac{1}{\mu}} + \lceil s_i \rfloor^{1-\frac{1}{\mu}} \right) \tag{5.22}$$

with $\mu > 1$ and $\rho \geq \delta$.

Remark 5.6 It is worth mentioning that the integral sliding surface (5.19) is invariant and insensitive to the observation errors $\tilde{x}_{0,k}^i$, if $\rho \geq \delta$ in protocol (5.22) is large enough. However, a larger ρ will result in a larger dithering. Thanks to the fixed-time convergence of (5.7), $\rho = \delta$ may be chosen for counteracting the unknown disturbances Δ_i merely.

The main result in this section is summarized in the following theorem.

Theorem 5.2 *Consider the multi-agent system described by* (5.1) *and* (5.2). *If Assumptions 5.1.1–5.1.3 hold, the fixed-time consensus tracking problem can be solved by the distributed protocol* (5.22) *with the decentralized observer* (5.4).

Proof The proof consists of three steps. We first show that all tracking errors $e_{i,k}$, $i \in \Gamma$, $k = 1, 2, \ldots, n$, are bounded on any time interval $[0, t]$, i.e., no finite-time escape occurs. Then we show that the reaching phase can be finished in a fixed-time under the protocol (5.22). Finally, we will prove that the tracking errors $e_{i,k}$ defined in (5.16) converge to zero in a fixed-time during the sliding motion $s_i = 0$.

(i) Consider the following Lyapunov function candidate:

$$V = \sum_{i=1}^{N} V_i = \sum_{i=1}^{N} |s_i|, \tag{5.23}$$

where $V_i = |s_i|$. Note that the definition of $\dot{V}$ becomes nontrivial when $s_i = 0$, since the right-hand side of (5.23) becomes discontinuous. The concept of Filippov solutions and set-valued Lie derivative needs to be applied [6]. Then one obtains

$$\dot{V} = \sum_{i=1}^{N} \mathrm{sign}(s_i)\dot{s}_i.$$

Taking into account (5.20) with (5.22), one gets

$$\begin{aligned}
\dot{V} &= \sum_{i=1}^{N} \mathrm{sign}(s_i)\Big(\dot{e}_{i,n} + \sum_{j=1}^{n} \big(k_i \lceil e_{i,j} \rfloor^{\gamma_i} + \bar{k}_i \lceil e_{i,j} \rfloor^{\bar{\gamma}_i}\big)\Big) \\
&= \sum_{i=1}^{N} k_i \left(-|s_i|^{1+\frac{1}{\mu}} - |s_i|^{1-\frac{1}{\mu}}\right) + \sum_{i=1}^{N} \mathrm{sign}(s_i)\left(\Delta_i - \rho\,\mathrm{sign}(s_i)\right) \\
&\quad - \alpha_n \sum_{i=1}^{N} \mathrm{sign}(s_i)\mathrm{sign}\left(\sum_{j=1}^{N} a_{ij}(\tilde{x}_{0,n}^{j} - \tilde{x}_{0,n}^{i}) - b_i \tilde{x}_{0,n}^{i}\right) \\
&\quad - \beta_n \sum_{i=1}^{N} \mathrm{sign}(s_i) \left\lceil \sum_{j=1}^{N} a_{ij}(\tilde{x}_{0,n}^{j} - \tilde{x}_{0,n}^{i}) - b_i \tilde{x}_{0,n}^{i} \right\rfloor^{2} \\
&\le -\sum_{i=1}^{N} k_i \left(|s_i|^{1+\frac{1}{\mu}} + |s_i|^{1-\frac{1}{\mu}}\right) \\
&\quad - \alpha_n \sum_{i=1}^{N} \mathrm{sign}(s_i)\mathrm{sign}\left(\sum_{j=1}^{N} a_{ij}(\tilde{x}_{0,n}^{j} - \tilde{x}_{0,n}^{i}) - b_i \tilde{x}_{0,n}^{i}\right) \\
&\quad - \beta_n \sum_{i=1}^{N} \mathrm{sign}(s_i) \left\lceil \sum_{j=1}^{N} a_{ij}(\tilde{x}_{0,n}^{j} - \tilde{x}_{0,n}^{i}) - b_i \tilde{x}_{0,n}^{i} \right\rfloor^{2}.
\end{aligned} \tag{5.24}$$

Recalling Theorem 5.1, it is shown that the estimate errors $\tilde{x}_{0,k}^i$ are fixed-time stable at the origin, which implies that $\tilde{x}_{0,k}^i$ are bounded, i.e., $|\tilde{x}_{0,k}^i| \leq \sigma_k^i \leq \bar{\sigma}$ for some constant $\bar{\sigma}$. Then there exists a constant C such that

$$\dot{V} \leq -\sum_{i=1}^{N} k_i |s_i| + C = -\underline{k} V + C, \tag{5.25}$$

with $\underline{k} = \min\{k_1, k_2, \ldots, k_N\}$. It follows that V is bounded on any time interval $[0, t]$, which, in turn, implies that s_i are bounded on any time interval. From (5.17) and the boundedness of all $\tilde{x}_{0,k}^i$, the finite-time escape of $e_{i,k+1}$ implies that $e_{i,k}$ will escape to infinity in finite time and have the same sign with $e_{i,k+1}$ ultimately, $k = 1, 2, \ldots, n-1$, $i \in \Gamma$, which leads to the unboundedness of s_i. This makes a contradiction. To this end, we can conclude that all tracking errors $e_{i,k}$ are bounded on any time interval $[0, t]$, i.e., no finite-time escape occurs.

(ii) For all $t \geq T_o$, the tracking error dynamics can be described by (5.18). Taking the time derivative of V in (5.23) and following the same line in (5.24) result in

$$\dot{V} \leq -\sum_{i=1}^{N} k_i \left(|s_i|^{1+\frac{1}{\mu}} + |s_i|^{1-\frac{1}{\mu}}\right). \tag{5.26}$$

It follows from Lemmas 3.1 and 3.2 that $\dot{V}$ in (5.26) satisfies

$$\begin{aligned} \dot{V} &\leq -N^{-\frac{1}{\mu}} \underline{k} \left(\sum_{i=1}^{N} |s_i|\right)^{1+\frac{1}{\mu}} - \underline{k} \left(\sum_{i=1}^{N} |s_i|\right)^{1-\frac{1}{\mu}} \\ &\leq -N^{-\frac{1}{\mu}} \underline{k} V^{1+\frac{1}{\mu}} - \underline{k} V^{1-\frac{1}{\mu}}. \end{aligned} \tag{5.27}$$

From Theorem 2.4, the system trajectory will reach the sliding surface $s_i = 0$ in a finite settling time bounded by $T_r \leq \pi \mu N^{1/2\mu}/(2\underline{k})$.

(iii) During the sliding motion (i.e., $s_i \equiv 0$), one gets that $\dot{s}_i = 0$ and the reduced closed-loop dynamics can be derived as, for $t \geq T_o + T_r$,

$$\begin{cases} \dot{e}_1(t) = e_2(t), \\ \dot{e}_2(t) = e_3(t), \\ \quad \vdots \\ \dot{e}_n(t) = -\sum_{i=1}^{n} \left(k_i \lceil e_i \rfloor^{\gamma_i} + \bar{k}_i \lceil e_i \rfloor^{\bar{\gamma}_i}\right). \end{cases} \tag{5.28}$$

It follows from Theorem 2.11 that the reduced dynamics (5.28) is fixed-time stable at the origin, i.e., there exists a constant T_s independent of initial conditions such that $e_{i,k} \to 0$ for all $t \geq T_o + T_r + T_s$, $i \in \Gamma$ and $k = 1, 2, \ldots, n$, where T_o is defined in Theorem 5.1. Note that $e_{i,k} \to 0$ implies that $x_{i,k} \to x_{0,k}$, when $\tilde{x}_{0,k}^i = 0$, which completes the proof. □

Remark 5.7 In practical implementation, one may assign $\hat{x}_0^i = x_0$ in (5.16) and (5.22) for the follower who has a direct access to the leader state, i.e., the ith follower for $b_i \neq 0$, to skip convergence phase of the observer, while the other followers with $b_i = 0$ use the reconstructed state $\hat{x}_i$.

5.3 A Numerical Example

In this section, an illustrative example is presented to show the efficacy of the proposed fixed-time consensus tracking protocol.

The multi-agent system consists of one leader indexed by 0 and five followers indexed by 1 to 5. The leader and the followers are described by, respectively,

$$\begin{cases} \dot{x}_{0,1}(t) = x_{0,2}(t), \\ \dot{x}_{0,2}(t) = x_{0,3}(t), \\ \dot{x}_{0,3}(t) = u_0(t), \end{cases} \tag{5.29}$$

and

$$\begin{cases} \dot{x}_{i,1}(t) = x_{i,2}(t), \\ \dot{x}_{i,2}(t) = x_{i,3}(t), \\ \dot{x}_{i,3}(t) = u_i(t) + \Delta_i(t), \quad (i = 1, 2, \ldots, 5), \end{cases} \tag{5.30}$$

where $x_{k,1}$, $x_{k,2}$, $x_{k,3}$, and u_k, $(k = 0, 1, 2, \ldots, 5)$, are the position, the velocity, the acceleration, and the input of the leader and the followers, respectively, Δ_i are the disturbances.

In simulation, the input of the leader is selected as $u_0 = -\sin(0.5t)$ and the external disturbances $\Delta_i = 0.1\sin(x_i)$ are used. The initial values of the leader are fixed as $x_0 = [-8, 0, 2]^T$. The interconnection topology is illustrated in Fig. 5.1 and the corresponding Laplacian matrix is

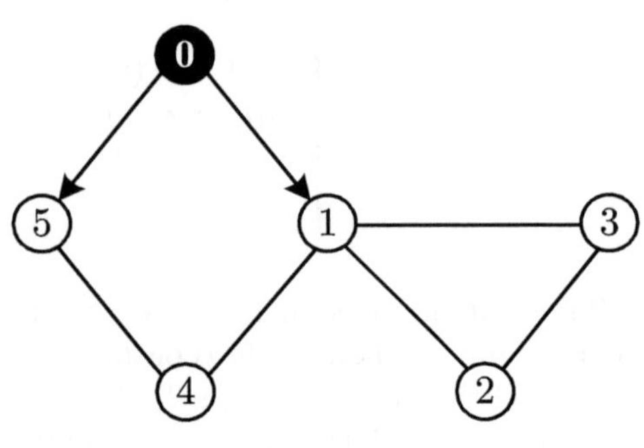

Fig. 5.1 Information flow among the leader and the followers

$$\mathscr{L} = \begin{bmatrix} 3 & -1 & -1 & -1 & 0 \\ -1 & 2 & -1 & 0 & 0 \\ -1 & -1 & 2 & 0 & 0 \\ -1 & 0 & 0 & 2 & -1 \\ 0 & 0 & 0 & -1 & 1 \end{bmatrix},$$

and the leader accessibility matrix is $\mathscr{B} = \text{diag}(1, 0, 0, 0, 1)$. The controller parameters $\gamma_1 = \frac{7}{16}, \gamma_2 = \frac{7}{13}, \gamma_3 = \frac{7}{10}, \bar{\gamma}_1 = \frac{13}{4}, \bar{\gamma}_2 = \frac{13}{7}, \bar{\gamma}_3 = \frac{13}{10}$, and $k_1 = \bar{k}_1 = 2$, $k_2 = \bar{k}_2 = 3$, $k_3 = \bar{k}_3 = 5$ are selected such that the resulting polynomial $s^3 + 5s^2 + 3s + 2$ is Hurwitz, i.e., its roots $\{-4.4241, -0.2880 \pm \text{j}0.6076\}$ are all located in the left complex plane. The observer parameters $\alpha_1 = \alpha_2 = 0.8$, $\alpha_3 = 2$ and $\beta_1 = \beta_2 = \beta_3 = 1.2$ are fixed in the simulation.

The simulation is carried out using Euler method with a fixed integration step equal to 10^{-4} [7]. In order to validate the efficacy of the proposed design, two scenarios for different initial values of the followers are considered:

(i) $x_1 = [-10, 0, 0]$, $x_2 = [-5, 0, 0]$, $x_3 = [0, 0, 0]$, $x_4 = [5, 0, 0]$, $x_5 = [10, 0, 0]$;

(ii) $x_1 = [-100, 0, 0]$, $x_2 = [-50, 0, 0]$, $x_3 = [10, 0, 0]$, $x_4 = [50, 0, 0]$, $x_5 = [100, 0, 0]$.

The simulation results for both scenarios are presented in Figs. 5.2, 5.3, 5.4, 5.5, 5.6, and 5.7. Figures 5.2, 5.3, and 5.4 show that the position, the velocity, and the acceleration errors between each follower and the leader converge to zero very fast, which implies that the consensus tracking is achieved in a fixed time. From Fig. 5.6, one can observe that the chattering happens in the control inputs of followers, which results from the discontinuous term $\rho\text{sign}(s_i)$ incorporated in the protocol (5.22).

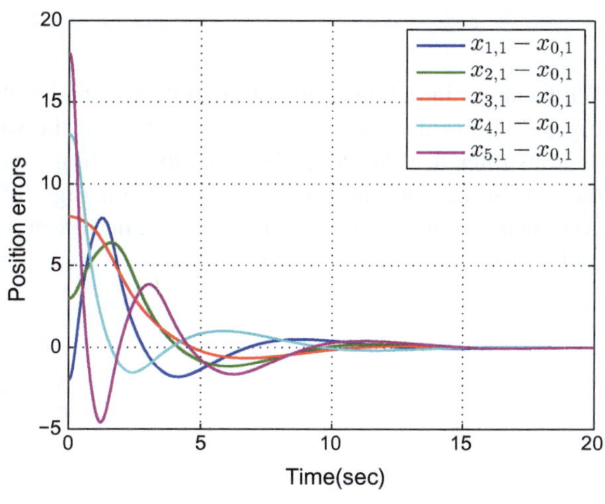

Fig. 5.2 Profiles of the position tracking errors: Scenario (i)

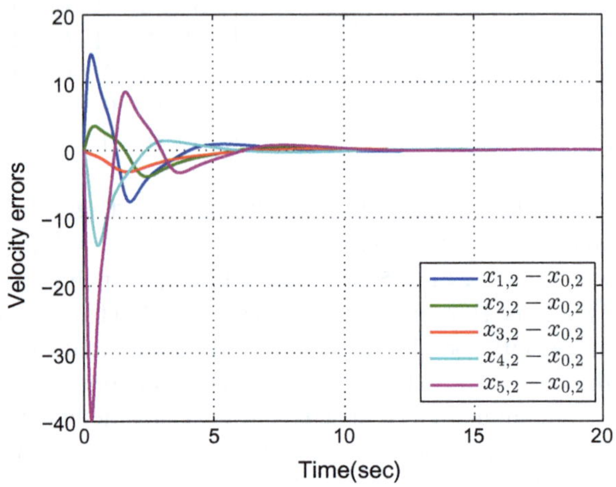

Fig. 5.3 Profiles of the velocity tracking errors: Scenario (i)

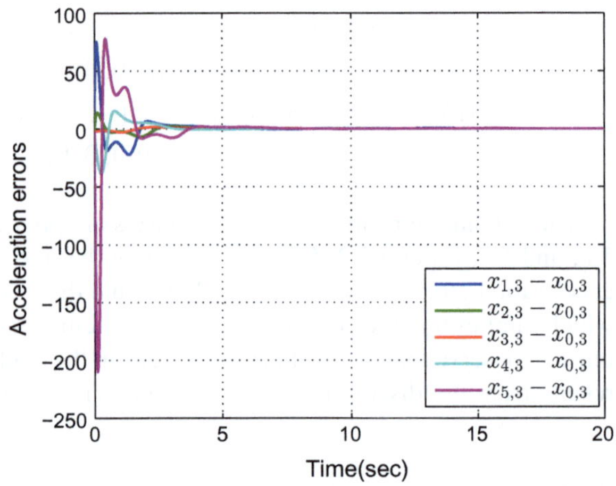

Fig. 5.4 Profiles of the acceleration tracking errors: Scenario (i)

With a larger initial deviation of the followers and without re-tuning the parameters, Fig. 5.7 shows the convergence time is almost the same as that in Fig. 5.2, which demonstrates a weak dependence of the settling time on initial conditions due to the fixed-time consensus design. Note from Figs. 5.5 and 5.8 that the distributed consensus observer achieves almost the same convergence speed to the leader's state under different initial scenarios.

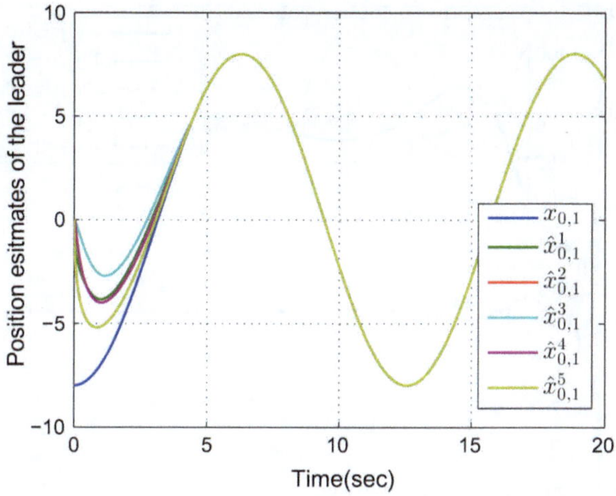

Fig. 5.5 Profiles of the estimates of the leader's position: Scenario (i)

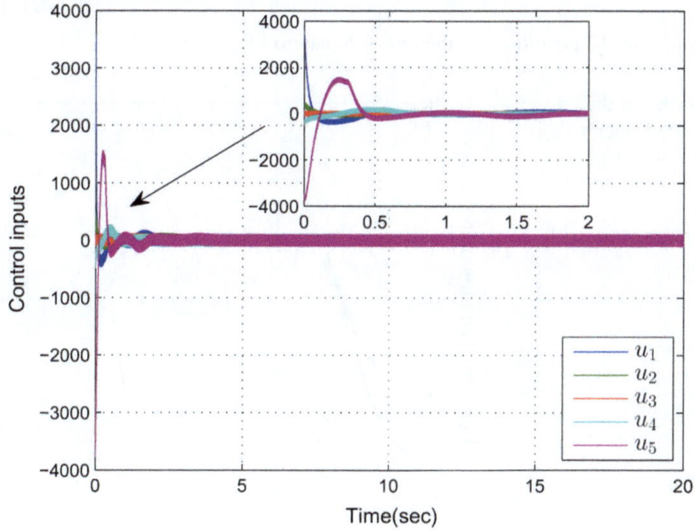

Fig. 5.6 Profiles of the control inputs of all followers: Scenario (i)

5.4 Conclusions

A fixed-time consensus tracking problem for multi-agent systems with high-order integrator dynamics is addressed. A distributed observer-based consensus tracking design framework is developed, which provides an effective way to construct the fixed-time consensus protocol for multi-agent systems with arbitrary order integrator dynamics.

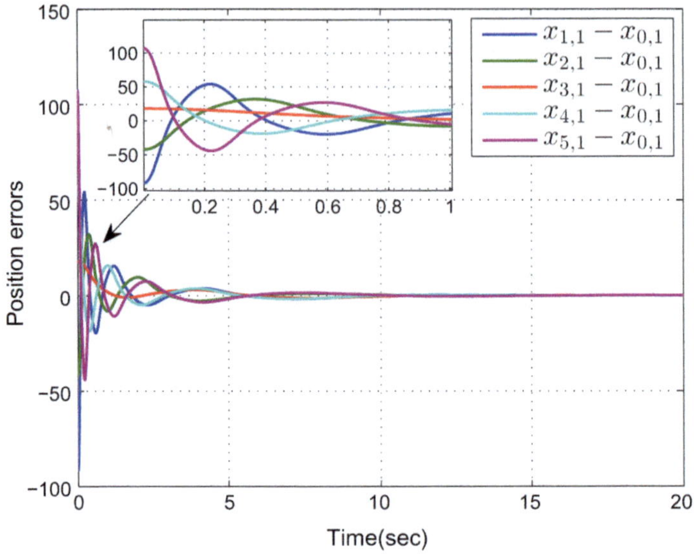

Fig. 5.7 Profiles of the position tracking errors: Scenario (ii)

Fig. 5.8 Profiles of the estimates of the leader's position: Scenario (ii)

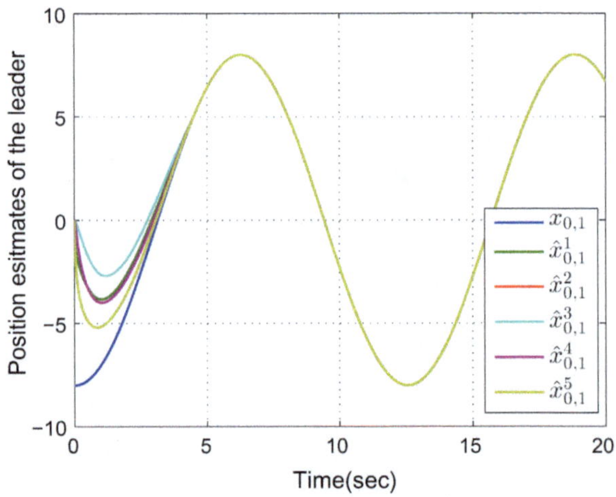

5.5 Notes

It is worth noting that the tracking protocol in [1, 8, 9] suffers from a communication loop issue due to the decoupling design, which prohibits the practical implementation. Based on the work in [10], the protocol proposed in this chapter completely overcomes this drawback due to the introduction of a distributed fixed-time observer to decouple the information interaction between the follower and the leader. For multivariable multi-agent systems with high-order dynamics, interested readers can refer to [11], where chattering reduction problem is dealt with in addition to fixed-time consensus tracking.

References

1. H. Zhang, F. Lewis, Adaptive cooperative tracking control of higher-order nonlinear systems with unknown dynamics. Automatica **48**(7), 1432–1439 (2012)
2. V. Utkin, *Sliding Modes in Control and Optimization* (Springer, Berlin, 1992)
3. A.S.M. Isira, Z. Zuo, Z. Ding, Leader-follower consensus control of Lipschitz nonlinear systems by output feedback. Int. J. Syst. Sci. **47**(16), 3772–3781 (2016)
4. Y. Cao, L. Zhang, C. Li, M.Z.Q. Chen, Observer-based consensus tracking of nonlinear agents in hybrid varying directed topology. IEEE Trans. Cybern. **47**(8), 2212–2222 (2017)
5. X. Liu, D.W.C. Ho, J. Cao, W. Xu, Discontinuous observers design for finite-time consensus of multiagent systems with external disturbances. IEEE Trans. Neural Netw. Learn. Syst. **28**(11), 2826–2830 (2017)
6. A. Filippov, *Differential Equations with Discontinuous Right-Hand Side* (Kluwer Academic, Dordrecht, 1988)
7. A. Levant, *On fixed and finte time stability in sliding mode control, in Proceedings of the IEEE Conference on Decision and Control* (Florence, Italy, 2013), pp. 4260–4265
8. S. Khoo, L. Xie, Z. Man, Robust finite-time consensus tracking algorithm for multirobot systems. IEEE/ASME Trans. Mechatron. **14**(2), 219–228 (2009)
9. Z. Zuo, Nonsingular fixed-time consensus tracking for second-order multi-agent networks. Automatica **54**, 305–309 (2015)
10. Z. Zuo, B. Tian, M. Defoort, Z. Ding, Fixed-time consensus tracking for multi-agent systems with high-order integrator dynamics. IEEE Trans. Autom. Control **63**(2), 563–570 (2018)
11. Z. Zuo, Q.-L. Han, B. Ning, An explicit estimate for the upper bound of the settling time in fixed-time leader-following consensus of high-order multi-variable multi-agent systems. IEEE Trans. Ind. Electron. **66**(8), 6250–6259 (2019)

Chapter 6
Fixed-Time Cooperative Control for Nonholonomic Chained-Form Multi-Agent Systems

This chapter deals with the consensus tracking problem for a multi-agent system with nonholonomic chained-form dynamics. A distributed observer is first proposed for each follower to estimate the state and the input of the leader in fixed-time under undirected communication graphs. Then, based on the observer and by adding a power integrator, a nonlinear protocol is designed such that the estimated leader state is tracked in fixed time. Different from some existing finite-time consensus tracking approaches, an explicit bound without dependence on initial states is derived for the settling time. Therefore, when initial conditions are unavailable, the proposed strategy is able to meet specific system requirements, e.g., a military target is tracked by a group of field robots in a prescribed time.

6.1 Problem Formulation

Consider a multi-agent system consisting of N followers indexed as agents $1, 2, \ldots,$ N and a leader indexed as agent 0. The dynamics of the ith follower, $i \in \Gamma$, is described by

$$\begin{cases} \dot{x}_{1,i}(t) = u_{1,i}(t), \\ \dot{x}_{2,i}(t) = x_{3,i}(t)u_{1,i}(t), \\ \quad\vdots \\ \dot{x}_{m-1,i}(t) = x_{m,i}(t)u_{1,i}(t), \\ \dot{x}_{m,i}(t) = u_{2,i}(t), \end{cases} \tag{6.1}$$

where $x_i(t) = [x_{1,i}, x_{2,i}, \ldots, x_{m,i}]^T \in \mathbb{R}^m$ is the ith agent's state, $u_i(t) = [u_{1,i}, u_{2,i}]^T$ $\in \mathbb{R}^2$ is the protocol to be designed. The dynamics of the leader is described by

Z. Zuo et al., *Fixed-Time Cooperative Control of Multi-Agent Systems*,
https://doi.org/10.1007/978-3-030-20279-8_6

$$\begin{cases} \dot{x}_{1,0}(t) = u_{1,0}(t), \\ \dot{x}_{2,0}(t) = x_{3,0}(t)u_{1,0}(t), \\ \quad \vdots \\ \dot{x}_{m-1,0}(t) = x_{m,0}(t)u_{1,0}(t), \\ \dot{x}_{m,0}(t) = u_{2,0}(t), \end{cases} \tag{6.2}$$

where $x_0(t) = [x_{1,0}, x_{2,0}, \ldots, x_{m,0}]^T \in \mathbb{R}^m$ is the leader's state, and $u_0(t) = [u_{1,0}, u_{2,0}]^T \in \mathbb{R}^2$ is the control input for the leader.

We first provide the definition of fixed-time consensus tracking for the system described by (6.1) and (6.2).

Definition 6.1 For the system described by (6.1) and (6.2), it is said to achieve fixed-time consensus tracking if there exists a suitable protocol $u_i(t)$ such that

$$\begin{cases} \lim_{t \to T} \|x_i(t) - x_0(t)\|_2 = 0; \\ \|x_i(t) - x_0(t)\|_2 = 0, \quad \forall t \geq T \end{cases}$$

for all $i \in \Gamma$, where T is upper bounded by $T_{\max}$ that is independent of initial states.

The problem to be addressed in this chapter is stated as: for the system described by (6.1) and (6.2), design a protocol $u_i(t)$ to achieve fixed-time consensus tracking.

Assumption 6.1.1 The input of the leader satisfies

$$|\dot{u}_{1,0}(t)| \leq c_1; \ |\dot{u}_{2,0}(t)| \leq c_2; \ c_3 \leq |u_{1,0}(t)| \leq \rho_1, \tag{6.3}$$

where c_1, c_2, c_3, and ρ_1 are known positive constants. Moreover, $u_{2,0}(t)$ is upper bounded. □

Assumption 6.1.2 $\mathscr{G}$ is undirected, and each follower can directly or indirectly obtain information from the leader. □

Remark 6.1 Assumption 6.1.1 indicates that the input (and its derivative) is not too large. For a wheeled mobile robot [1], Assumption 6.1.1 implies that the linear acceleration, the angular acceleration, the linear velocity, and the angular velocity are all bounded, which is a realistic consideration.

In what follows, we occasionally drop the dependence on t of states for notational simplicity.

Lemma 6.1 ([2]) *Let x and y be real numbers. If $0 < p \leq 1$ and is a ratio of two positive odd integers, then $|x^p - y^p| \leq 2^{1-p}|x - y|^p$.*

Lemma 6.2 ([2]) *Let c and d be positive real numbers, and $\varsigma(x, y) > 0$ be a real-valued function. Then, $|x|^c|y|^d \leq c\varsigma(x, y)|x|^{c+d}/(c + d) + d\varsigma^{-c/d}(x, y)|y|^{c+d}/(c + d)$.*

Proof The inequality is deduced from the Young's inequality

$$ab \leq \frac{a^r}{r} + \frac{(r-1)b^{r/(r-1)}}{r}, r > 1, a \geq 0, b \geq 0$$

with

$$a = |x|^c \varsigma^{c/(c+d)}, \quad b = |y|^d \varsigma^{-c/(c+d)}, \quad r = \frac{c+d}{c}.$$

6.2 Fixed-Time Consensus Tracking

Distributed observers are proposed to estimate the leader's state and the leader's input under undirected graphs in fixed time. Then, an observer-based nonlinear protocol is designed such that the estimated leader state is tracked in fixed time.

6.2.1 Distributed Observer

We propose a distributed observer for the ith follower, i.e., $i = 1, 2, \ldots, N$,

$$\begin{cases} \dot{\hat{x}}_{1,i} = -\alpha \left(\sum_{j=1}^{N} a_{ij}(\hat{x}_{1,i} - \hat{x}_{1,j}) + b_i(\hat{x}_{1,i} - x_{1,0}) \right)^{1-\frac{a}{b}} \\ \qquad -\beta \left(\sum_{j=1}^{N} a_{ij}(\hat{x}_{1,i} - \hat{x}_{1,j}) + b_i(\hat{x}_{1,i} - x_{1,0}) \right)^{1+\frac{a}{b}} + \hat{u}_{1,i}, \\ \dot{\hat{x}}_{k,i} = -\alpha \left(\sum_{j=1}^{N} a_{ij}(\hat{x}_{k,i} - \hat{x}_{k,j}) + b_i(\hat{x}_{k,i} - x_{k,0}) \right)^{1-\frac{a}{b}} \\ \qquad -\beta \left(\sum_{j=1}^{N} a_{ij}(\hat{x}_{k,i} - \hat{x}_{k,j}) + b_i(\hat{x}_{k,i} - x_{k,0}) \right)^{1+\frac{a}{b}} + \hat{x}_{k+1,i}\hat{u}_{1,i}, \\ \qquad (k = 2, 3, \ldots, m-1), \\ \dot{\hat{x}}_{m,i} = -\alpha \left(\sum_{j=1}^{N} a_{ij}(\hat{x}_{m,i} - \hat{x}_{m,j}) + b_i(\hat{x}_{m,i} - x_{m,0}) \right)^{1-\frac{a}{b}} \\ \qquad -\beta \left(\sum_{j=1}^{N} a_{ij}(\hat{x}_{m,i} - \hat{x}_{m,j}) + b_i(\hat{x}_{m,i} - x_{m,0}) \right)^{1+\frac{a}{b}} + \hat{u}_{2,i}, \end{cases} \tag{6.4}$$

and for $l = 1, 2$, $i = 1, 2, \ldots, N$,

$$
\begin{aligned}
\dot{\hat{u}}_{l,i} = & -\zeta_l \operatorname{sign}\left(\sum_{j=1}^{N} a_{ij}(\hat{u}_{l,i} - \hat{u}_{l,j}) + b_i(\hat{u}_{l,i} - u_{l,0})\right) \\
& - \gamma \left(\sum_{j=1}^{N} a_{ij}(\hat{u}_{l,i} - \hat{u}_{l,j}) + b_i(\hat{u}_{l,i} - u_{l,0})\right)^{1-\frac{a}{b}} \\
& - \chi \left(\sum_{j=1}^{N} a_{ij}(\hat{u}_{l,i} - \hat{u}_{l,j}) + b_i(\hat{u}_{l,i} - u_{l,0})\right)^{1+\frac{a}{b}}, \qquad (6.5)
\end{aligned}
$$

where $\hat{x}_i = [\hat{x}_{1,i}, \hat{x}_{2,i}, \ldots, \hat{x}_{m,i}]^T \in \mathbb{R}^m$ and $\hat{u}_i = [\hat{u}_{1,i}, \hat{u}_{2,i}]^T \in \mathbb{R}^2$ are the ith follower's estimate of the leader's state x_0 and the leader's input u_0, respectively; a and b satisfying $a < b$ are a positive even integer and a positive odd integer, respectively; α, β, γ, χ, and ζ_l, $l = 1, 2$, are observer gains satisfying

$$
\begin{cases}
\alpha = \dfrac{\alpha_0}{\left(2\lambda_{\min}(\mathcal{M})\right)^{1-\frac{a}{2b}}}, \ \beta = \dfrac{\beta_0 N^{\frac{a}{2b}}}{\left(2\lambda_{\min}(\mathcal{M})\right)^{1+\frac{a}{2b}}}, \\
\gamma = \dfrac{\gamma_0}{\left(2\lambda_{\min}(\mathcal{M})\right)^{1-\frac{a}{2b}}}, \ \chi = \dfrac{\chi_0 N^{\frac{a}{2b}}}{\left(2\lambda_{\min}(\mathcal{M})\right)^{1+\frac{a}{2b}}}, \\
\zeta_l \geq c_l, \quad l = 1, 2,
\end{cases} \qquad (6.6)
$$

where α_0, β_0, γ_0, and χ_0 are preselected positive constants, and $\lambda_{\min}(\cdot)$ denotes the smallest eigenvalue of a matrix. We now present the following result.

Theorem 6.1 *Under Assumptions 6.1.1 and 6.1.2, for the system described by (6.1) and (6.2), by using the observer described by (6.4)–(6.6), each follower accurately estimates the leader's state and the leader's input in a finite-time T, i.e., $\lim_{t\to T} \|\hat{x}_i(t) - x_0(t)\|_2 = 0$ and $\lim_{t\to T} \|\hat{u}_i(t) - u_0(t)\|_2 = 0$, $\forall i \in \Gamma$, $\|\hat{x}_i(t) - x_0(t)\|_2 = 0$ and $\|\hat{u}_i(t) - u_0(t)\|_2 = 0$ for $t \geq T$. Moreover, T is upper bounded by T_{max1}, i.e.,*

$$
T \leq T_{max1} := \frac{\pi b}{a\sqrt{\gamma_0 \chi_0}} + \frac{\pi b(m-1)}{a\sqrt{\alpha_0 \beta_0}}. \qquad (6.7)
$$

Proof Let $\tilde{x}_i = [\tilde{x}_{1,i}, \tilde{x}_{2,i}, \ldots, \tilde{x}_{m,i}]^T = \hat{x}_i - x_0 = [\hat{x}_{1,i} - x_{1,0}, \hat{x}_{2,i} - x_{2,0}, \ldots, \hat{x}_{m,i} - x_{m,0}]^T$, $\forall i \in \Gamma$, and $\tilde{u}_i = [\tilde{u}_{1,i}, \tilde{u}_{2,i}]^T = \hat{u}_i - u_0 = [\hat{u}_{1,i} - u_{1,0}, \hat{u}_{2,i} - u_{2,0}]^T$.

(i) We first prove that each follower can accurately estimate the leader's input in fixed time. From (6.5), for $l = 2$, one obtains the time derivative of $\tilde{u}_{2,i}$ as

$$
\begin{aligned}
\dot{\tilde{u}}_{2,i} =& -\dot{u}_{2,0} - \gamma\left(\sum_{j=1}^{N} a_{ij}(\tilde{u}_{2,i} - \tilde{u}_{2,j}) + b_i\tilde{u}_{2,i}\right)^{1-\frac{a}{b}} \\
& - \chi\left(\sum_{j=1}^{N} a_{ij}(\tilde{u}_{2,i} - \tilde{u}_{2,j}) + b_i\tilde{u}_{2,i}\right)^{1+\frac{a}{b}} \\
& - \zeta_2\text{sign}\left(\sum_{j=1}^{N} a_{ij}(\tilde{u}_{2,i} - \tilde{u}_{2,j}) + b_i\tilde{u}_{2,i}\right).
\end{aligned}
\tag{6.8}
$$

Let $\tilde{U}_2 = [\tilde{u}_{2,1}, \tilde{u}_{2,2}, \ldots, \tilde{u}_{2,N}]^T$. Then, (6.8) can be written in a compact vector form of $\dot{\tilde{U}}_2 = -\gamma(\mathscr{M}\tilde{U}_2)^{1-\frac{a}{b}} - \chi(\mathscr{M}\tilde{U}_2)^{1+\frac{a}{b}} - \zeta_2\text{sign}(\mathscr{M}\tilde{U}_2) - \dot{u}_{2,0}\mathbf{1}_N$. Construct a Lyapunov function of $V_1 = \frac{1}{2}\tilde{U}_2^T\mathscr{M}\tilde{U}_2$, whose time derivative is

$$
\begin{aligned}
\dot{V}_1 =& \tilde{U}_2^T\mathscr{M}\left(-\gamma(\mathscr{M}\tilde{U}_2)^{1-\frac{a}{b}} - \chi(\mathscr{M}\tilde{U}_2)^{1+\frac{a}{b}} - \zeta_2\text{sign}(\mathscr{M}\tilde{U}_2) - \dot{u}_{2,0}\mathbf{1}_N\right) \\
\le& -\gamma\|\mathscr{M}\tilde{U}_2\|_{2-\frac{a}{b}}^{2-\frac{a}{b}} - \chi\|\mathscr{M}\tilde{U}_2\|_{2+\frac{a}{b}}^{2+\frac{a}{b}} - (\zeta_2 - c_2)\|\mathscr{M}\tilde{U}_2\|_1 \\
\le& -\gamma(2V_1\lambda_{\min}(\mathscr{M}))^{1-\frac{a}{2b}} - \chi N^{-\frac{a}{2b}}(2V_1\lambda_{\min}(\mathscr{M}))^{1+\frac{a}{2b}} \\
=& -\gamma_0 V_1^{1-\frac{a}{2b}} - \chi_0 V_1^{1+\frac{a}{2b}},
\end{aligned}
\tag{6.9}
$$

where the fact that $\|\mathscr{M}\tilde{U}_2\|_{2-\frac{a}{b}} \ge \|\mathscr{M}\tilde{U}_2\|_2$, $\|\mathscr{M}\tilde{U}_2\|_{2+\frac{a}{b}} \ge N^{-\frac{a}{2a+4b}}\|\mathscr{M}\tilde{U}_2\|_2$, and $\|\mathscr{M}\tilde{U}_2\|_2^2 \ge 2V_1\lambda_{\min}(\mathscr{M})$ [3] is used. By using Theorem 2.4 and *Comparison Principle*, V_1 converges to zero with settling time bounded by $T_1 := \frac{\pi b}{a\sqrt{\gamma_0\chi_0}}$, which is independent of initial states. Note that $V_1 = 0$ indicates that $\tilde{u}_{2,i} = 0$, $i \in \varGamma$. Following a similar procedure above, it can be concluded that $\tilde{u}_{1,i} = 0$ within T_1, $i \in \varGamma$.

(ii) We then prove that each follower can accurately estimate the leader's state in fixed time. Since $\tilde{x}_{m,i} = \hat{x}_{m,i} - x_{m,0}$, one obtains that $\dot{\tilde{x}}_{m,i} = \hat{u}_{2,i} - u_{2,0} - \alpha(\sum_{j=1}^{N} a_{ij}(\tilde{x}_{m,i} - \tilde{x}_{m,j}) + b_i\tilde{x}_{m,i})^{1-\frac{a}{b}} - \beta(\sum_{j=1}^{N} a_{ij}(\tilde{x}_{m,i} - \tilde{x}_{m,j}) + b_i\tilde{x}_{m,i})^{1+\frac{a}{b}}$. When $t \ge T_1$, from the analysis in (i), one obtains that $\hat{u}_{2,i} - u_{2,0} = 0$.

Let $\tilde{X}_m = [\tilde{x}_{m,1}, \tilde{x}_{m,2}, \ldots, \tilde{x}_{m,N}]^T$. Then, the dynamics of $\tilde{x}_{m,i}$ can be written in a compact vector form of $\dot{\tilde{X}}_m = -\alpha(\mathscr{M}\tilde{X}_m)^{1-\frac{a}{b}} - \beta(\mathscr{M}\tilde{X}_m)^{1-\frac{a}{b}}$. Construct a Lyapunov function of $V_m = \frac{1}{2}\tilde{X}_m^T\mathscr{M}\tilde{X}_m$, whose time derivative is

$$
\begin{aligned}
\dot{V}_m =& \tilde{X}_m^T\mathscr{M}\left(-\alpha(\mathscr{M}\tilde{X}_m)^{1-\frac{a}{b}} - \beta(\mathscr{M}\tilde{X}_m)^{1+\frac{a}{b}}\right) \\
\le& -\alpha\|\mathscr{M}\tilde{X}_m\|_{2-\frac{a}{b}}^{2-\frac{a}{b}} - \beta\|\mathscr{M}\tilde{X}_m\|_{2+\frac{a}{b}}^{2+\frac{a}{b}} \\
\le& -\alpha(2V_m\lambda_{\min}(\mathscr{M}))^{1-\frac{a}{2b}} - \beta N^{-\frac{a}{2b}}(2V_m\lambda_{\min}(\mathscr{M}))^{1+\frac{a}{2b}} \\
=& -\alpha_0 V_m^{1-\frac{a}{2b}} - \beta_0 V_m^{1+\frac{a}{2b}}.
\end{aligned}
$$

By using Theorem 2.4 and *Comparison Principle*, $V_m = 0$ converges to zero with settling time bounded by $T_2 := \frac{\pi b}{a\sqrt{\alpha_0\beta_0}}$. $V_m = 0$ indicates that $\tilde{x}_{m,i} = 0$, $i \in \Gamma$. One obtains the dynamics of $\tilde{x}_{m-1,i}$ as

$$\begin{aligned}\dot{\tilde{x}}_{m-1,i} = & -\alpha\left(\sum_{j=1}^{N} a_{ij}(\tilde{x}_{m-1,i} - \tilde{x}_{m-1,j}) + b_i\tilde{x}_{m-1,i}\right)^{1-\frac{a}{b}} \\ & - \beta\left(\sum_{j=1}^{N} a_{ij}(\tilde{x}_{m-1,i} - \tilde{x}_{m-1,j}) + b_i\tilde{x}_{m-1,i}\right)^{1+\frac{a}{b}} \\ & + \hat{x}_{m,i}\hat{u}_{1,i} - x_{m,0}u_{1,0}.\end{aligned}$$

When $t \geq T_1 + T_2$, $\hat{x}_{m,i}\hat{u}_{1,i} - x_{m,0}u_{1,0} = 0$. Let $\tilde{X}_{m-1} = [\tilde{x}_{m-1,1}, \tilde{x}_{m-1,2}, \ldots, \tilde{x}_{m-1,N}]^T$. A Lyapunov function $V_{m-1} = \frac{1}{2}\tilde{X}_{m-1}^T \mathscr{M} \tilde{X}_{m-1}$ can be accordingly constructed. Following a recursive manner, it can be proved that $\tilde{x}_{2,i}$ converges to zero in a finite settling time bounded by $T_{\max 1} = T_1 + (m-1)T_2 = \frac{\pi b}{a\sqrt{\gamma_0\chi_0}} + \frac{\pi b(m-1)}{a\sqrt{\alpha_0\beta_0}}$. Besides, $\tilde{x}_{1,i}$ converges to zero within $T_1 + T_2$. Therefore, all followers accurately estimate the leader's state and the leader's input within $\frac{\pi b}{a\sqrt{\gamma_0\chi_0}} + \frac{\pi b(m-1)}{a\sqrt{\alpha_0\beta_0}}$. □

6.2.2 *Consensus Tracking*

In this subsection, observer-based nonlinear protocols are proposed to achieve consensus tracking in fixed time under undirected graphs. First, the tracking error is defined by $e_{k,i} = x_{k,i} - \hat{x}_{k,i}$, $k = 1, 2, \ldots, m$ and $i = 1, 2, \ldots, N$. Combining (6.1) and (6.4), the time derivative of $e_{k,i}$ is

$$\begin{cases} \dot{e}_{1,i} = u_{1,i} - \hat{u}_{1,i} + \alpha\left(\sum_{j=1}^{N} a_{ij}(\tilde{x}_{1,i} - \tilde{x}_{1,j}) + b_i\tilde{x}_{1,i}\right)^{1-\frac{a}{b}} \\ \qquad +\beta\left(\sum_{j=1}^{N} a_{ij}(\tilde{x}_{1,i} - \tilde{x}_{1,j}) + b_i\tilde{x}_{1,i}\right)^{1+\frac{a}{b}}, \\ \dot{e}_{k,i} = u_{1,i}x_{k+1,i} + \alpha\left(\sum_{j=1}^{N} a_{ij}(\tilde{x}_{k,i} - \tilde{x}_{k,j}) + b_i\tilde{x}_{k,i}\right)^{1-\frac{a}{b}} \\ \qquad -\hat{u}_{1,i}\hat{x}_{k+1,i} + \beta\left(\sum_{j=1}^{N} a_{ij}(\tilde{x}_{k,i} - \tilde{x}_{k,j}) + b_i\tilde{x}_{k,i}\right)^{1+\frac{a}{b}}, \\ \qquad k = 2, 3, \ldots, m-1, \\ \dot{e}_{m,i} = u_{2,i} - \hat{u}_{2,i} + \alpha\left(\sum_{j=1}^{N} a_{ij}(\tilde{x}_{m,i} - \tilde{x}_{m,j}) + b_i\tilde{x}_{m,i}\right)^{1-\frac{a}{b}} \\ \qquad +\beta\left(\sum_{j=1}^{N} a_{ij}(\tilde{x}_{m,i} - \tilde{x}_{m,j}) + b_i\tilde{x}_{m,i}\right)^{1+\frac{a}{b}}. \end{cases} \tag{6.10}$$

In Theorem 6.1, $k \in \{1, 2, \ldots, m\}$, $i \in \Gamma$, it is proved that $\tilde{x}_{k,i} = 0$ when $t \geq T_{\max 1}$. Therefore, when $t \geq T_{\max 1}$, (6.10) can be rewritten as

$$\begin{cases} \dot{e}_{1,i} = u_{1,i} - \hat{u}_{1,i}, \\ \dot{e}_{k,i} = u_{1,i}x_{k+1,i} - \hat{u}_{1,i}\hat{x}_{k+1,i}, \quad k = 2, 3, \ldots, m-1, \\ \dot{e}_{m,i} = u_{2,i} - \hat{u}_{2,i}. \end{cases} \tag{6.11}$$

In what follows, our aim is to design tracking protocols $u_{1,i}$ and $u_{2,i}$ in (6.11) for each follower, $i \in \Gamma$, such that the estimated state of the leader is tracked in a fixed time. The proof consists of two parts: the stabilization of $e_{1,i}$ and the stabilization of $e_{2,i}, e_{3,i}, \ldots, e_{m,i}$.

First, a tracking protocol $u_{1,i}$ is proposed

$$u_{1,i} = \hat{u}_{1,i} - 2^{-1+\frac{a}{2b}}\alpha_0 e_{1,i}^{1-\frac{a}{b}} - 2^{-1-\frac{a}{2b}}\beta_0 e_{1,i}^{1+\frac{a}{b}}. \tag{6.12}$$

By using Theorem 2.4, one obtains that $e_{1,i} = 0$ is fixed-time stable within $T_{\max 1} + T_2$. When $t \geq T_{\max 1} + T_2$, one obtains that $u_{1,i} = \hat{u}_{1,i}$. Therefore, the last $m-1$ equations in (6.11) are rewritten as

$$\begin{cases} \dot{e}_{k,i} = \hat{u}_{1,i}e_{k+1,i}, \quad k = 2, 3, \ldots, m-1, \\ \dot{e}_{m,i} = u_{2,i} - \hat{u}_{2,i}. \end{cases} \tag{6.13}$$

What remains is to design a tracking protocol $u_{2,i}$ such that (6.13) is fixed-time stable. Introduce the notation of

$$q_\iota = 1 + (\iota - 1)\sigma, \quad \iota = 1, 2, \ldots, m, \tag{6.14}$$

where $\sigma = -v/w$ with v and w being a positive even integer and a positive odd integer, respectively. In addition, it is required that $0 < q_\iota < 1$, for $\iota = 2, 3, \ldots, m$, so $\sigma \in (-1/(m-1), 0)$.

We now propose a nonlinear protocol for the ith follower,

$$u_{2,i} = \hat{u}_{2,i} - |\hat{u}_{1,i}|\xi_{m-1,i}^{q_m}\big(1 + r\xi_{m-1,i}^{\kappa} + g_{m-2}\hat{F}_{m-2,i}(e_{2,i}, \ldots, e_{m-1,i})\big), \tag{6.15}$$

where $r > 0$, $\kappa = 2 - 2q_2$, $\xi_{m-1,i}$ is a virtual protocol to be designed in a recursive manner, $g_{m-2} > 0$ is a sufficiently large constant, $\hat{F}_{m-2,i}(\cdot) > 0$ is a properly selected C^1 function. Obviously, κ is the ratio of a positive even integer and a positive odd integer. The design procedure is integrated in the following theorem.

Theorem 6.2 *Under Assumptions 6.1.1 and 6.1.2, for the system described by (6.1) and (6.2), the fixed-time consensus tracking is achieved by using the protocols (6.12)–(6.15) together with the observer described by (6.4)–(6.6), i.e.,* $\lim_{t\to T}|x_{k,i}(t) - x_{k,0}(t)| = 0$, $\forall i \in \Gamma$, $k = 1, 2, \ldots, m$, *and* $|x_{k,i}(t) - x_{k,0}(t)| = 0$ *for* $t \geq T$. *Moreover,* T *is upper bounded by a fixed time, i.e.,*

$$T \leq T_{max1} + T_{max2} + T_2, \tag{6.16}$$

where

$$T_{max2} := \frac{2\pi}{c_3(1-q_2)\sqrt{r}(m-1)^{\frac{q_2-1}{4}}}. \tag{6.17}$$

Proof The tracking protocol for (6.13) is designed in a recursive manner.
Step 1: Construct the Lyapunov function $V_{2,i} = \frac{1}{2}e_{2,i}^2$, whose time derivative is

$$\dot{V}_{2,i} = \hat{u}_{1,i}e_{2,i}e_{3,i} = \hat{u}_{1,i}e_{2,i}e_{3,i}^* + \hat{u}_{1,i}e_{2,i}(e_{3,i} - e_{3,i}^*). \tag{6.18}$$

Design a virtual protocol as

$$\begin{aligned} e_{3,i}^* &= -\text{sign}(\hat{u}_{1,i})e_{2,i}^{q_2}(m - 1 + re_{2,i}^{\kappa}) \\ &\triangleq -\text{sign}(\hat{u}_{1,i})\xi_{1,i}^{q_2}F_{1,i}(e_{2,i}), \end{aligned} \tag{6.19}$$

where $\xi_{1,i} = e_{2,i}$, so (6.18) can be rewritten as

$$\dot{V}_{2,i} = -(m-1)|\hat{u}_{1,i}|\xi_{1,i}^{q_2+1} - r|\hat{u}_{1,i}|\xi_{1,i}^{q_2+\kappa+1} + \hat{u}_{1,i}\xi_{1,i}(e_{3,i} - e_{3,i}^*).$$

Step 2: Let $\xi_{2,i} = e_{3,i}^{1/q_2} - e_{3,i}^{*\ 1/q_2}$, and construct the following Lyapunov function $V_{3,i} = V_{2,i} + \int_{e_{3,i}^*}^{e_{3,i}} (s^{1/q_2} - e_{3,i}^{*\ 1/q_2})^{2-q_2}\, \mathrm{d}s$, whose time derivative is

$$\begin{aligned} \dot{V}_{3,i} =& -(m-1)|\hat{u}_{1,i}|\xi_{1,i}^{q_2+1} - r|\hat{u}_{1,i}|\xi_{1,i}^{q_2+\kappa+1} + \hat{u}_{1,i}\xi_{1,i}(e_{3,i} - e_{3,i}^*) + \xi_{2,i}^{2-q_2}\hat{u}_{1,i}e_{4,i} \\ &- (2-q_2)\frac{\mathrm{d}e_{3,i}^{*\ \frac{1}{q_2}}}{\mathrm{d}t}\int_{e_{3,i}^*}^{e_{3,i}} (s^{\frac{1}{q_2}} - e_{3,i}^{*\ \frac{1}{q_2}})^{1-q_2}\, \mathrm{d}s. \end{aligned} \tag{6.20}$$

For the third term in (6.20), by using Lemmas 6.1 and 6.2, one obtains

$$\begin{aligned} \hat{u}_{1,i}\xi_{1,i}(e_{3,i} - e_{3,i}^*) &\le |\hat{u}_{1,i}||\xi_{1,i}|\left|(e_{3,i}^{\frac{1}{q_2}})^{q_2} - (e_{3,i}^{*\ \frac{1}{q_2}})^{q_2}\right| \\ &\le 2^{1-q_2}|\hat{u}_{1,i}||\xi_{1,i}||\xi_{2,i}|^{q_2} \\ &\le |\hat{u}_{1,i}|\Big(\frac{1}{2}\xi_{1,i}^{q_2+1} + g_1\xi_{2,i}^{q_2+1}\Big) \end{aligned} \tag{6.21}$$

for a positive constant g_1. Next, from (6.19), one obtains

$$\begin{aligned}
\left|\frac{\mathrm{d}e_{3,i}^{*\frac{1}{q_2}}}{\mathrm{d}t}\right| &\le \left(F_{1,i}^{\frac{1}{q_2}}(e_{2,i}) + |e_{2,i}|\left|\frac{\mathrm{d}F_{1,i}^{\frac{1}{q_2}}(e_{2,i})}{\mathrm{d}e_{2,i}}\right|\right)|\hat{u}_{1,i}e_{3,i}| \\
&\le \left(F_{1,i}^{\frac{1}{q_2}}(e_{2,i}) + |e_{2,i}|\left|\frac{\mathrm{d}F_{1,i}^{\frac{1}{q_2}}(e_{2,i})}{\mathrm{d}e_{2,i}}\right|\right) \\
&\quad \times |\hat{u}_{1,i}|\big(|\xi_{2,i}|^{q_2} + |\xi_{1,i}|^{q_2}F_{1,i}(e_{2,i})\big) \\
&\le |\hat{u}_{1,i}|\big(|\xi_{1,i}|^{q_2} + |\xi_{2,i}|^{q_2}\big)\tilde{F}_{1,i}(e_{2,i}) \qquad (6.22)
\end{aligned}$$

with $\tilde{F}_{1,i}(e_{2,i}) > 0$ being a C^1 function. Since $e_{3,i} = (\xi_{2,i} + e_{3,i}^{*\ 1/q_2})^{q_2}$, one obtains that $|e_{3,i}| \le (|\xi_{2,i}| + |e_{3,i}^*|^{1/q_2})^{q_2} \le |\xi_{2,i}|^{q_2} + |e_{3,i}^*| = |\xi_{2,i}|^{q_2} + |\xi_{1,i}|^{q_2}F_{1,i}(e_{2,i})$, which is used to obtain the second inequality in (6.22). For the last term in (6.20), one obtains

$$\begin{aligned}
&\left|(2-q_2)\frac{\mathrm{d}e_{3,i}^{*\frac{1}{q_2}}}{\mathrm{d}t}\int_{e_{3,i}^*}^{e_{3,i}}(s^{\frac{1}{q_2}} - e_{3,i}^{*\frac{1}{q_2}})^{1-q_2}\,\mathrm{d}s\right| \\
&\quad \le (2-q_2)|\hat{u}_{1,i}|\big(|\xi_{1,i}|^{q_2} + |\xi_{2,i}|^{q_2}\big)\tilde{F}_{1,i}(e_{2,i})|e_{3,i} - e_{3,i}^*||\xi_{2,i}|^{1-q_2} \\
&\quad \le 2^{1-q_2}(2-q_2)|\hat{u}_{1,i}||\xi_{2,i}|\big(|\xi_{1,i}|^{q_2} + |\xi_{2,i}|^{q_2}\big)\tilde{F}_{1,i}(e_{2,i}) \\
&\quad \le |\hat{u}_{1,i}|\left(\frac{1}{2}\xi_{1,i}^{q_2+1} + \xi_{2,i}^{q_2+1}\hat{F}_{1,i}(e_{2,i})\right) \qquad (6.23)
\end{aligned}$$

with $\hat{F}_{1,i}(e_{2,i}) > 0$ being a C^1 function. Combining (6.20), (6.21), and (6.23) yields

$$\begin{aligned}
\dot{V}_{3,i} \le& -(m-2)|\hat{u}_{1,i}|\xi_{1,i}^{q_2+1} - r|\hat{u}_{1,i}|\xi_{1,i}^{q_2+\kappa+1} + \xi_{2,i}^{2-q_2}\hat{u}_{1,i}(e_{4,i} - e_{4,i}^*) \\
&+ \xi_{2,i}^{2-q_2}\hat{u}_{1,i}e_{4,i}^* + |\hat{u}_{1,i}|\xi_{2,i}^{q_2+1}\big(g_1 + \hat{F}_{1,i}(e_{2,i})\big).
\end{aligned}$$

Design a virtual protocol as

$$\begin{aligned}
e_{4,i}^* =& -\operatorname{sign}(\hat{u}_{1,i})\xi_{2,i}^{q_3}\big(m - 2 + r\xi_{2,i}^{\kappa} + g_1 + \hat{F}_{1,i}(e_{2,i})\big) \\
\triangleq& -\operatorname{sign}(\hat{u}_{1,i})\xi_{2,i}^{q_3}F_{2,i}(e_{2,i}, e_{3,i}). \qquad (6.24)
\end{aligned}$$

Then, one gets

$$\begin{aligned}
\dot{V}_{3,i} \le& -(m-2)|\hat{u}_{1,i}|(\xi_{1,i}^{q_2+1} + \xi_{2,i}^{q_2+1}) \\
&- r|\hat{u}_{1,i}|(\xi_{1,i}^{q_2+\kappa+1} + \xi_{2,i}^{q_2+\kappa+1}) + \xi_{2,i}^{2-q_2}\hat{u}_{1,i}(e_{4,i} - e_{4,i}^*).
\end{aligned}$$

Inductive Step: For $i = 1, 2, \ldots, N$, at Step $k-1$, suppose that there exist a set of virtual protocols $e^*_{2,i}, e^*_{3,i}, \ldots, e^*_{k,i}$, i.e.,

$$
\begin{aligned}
e^*_{2,i} &= 0, & \xi_{1,i} &= e_{2,i}^{\frac{1}{q_1}} - {e^*_{2,i}}^{\frac{1}{q_1}} \\
e^*_{3,i} &= -\operatorname{sign}(\hat{u}_{1,i})\xi_{1,i}^{q_2} F_{1,i}(\cdot), & \xi_{2,i} &= e_{3,i}^{\frac{1}{q_2}} - {e^*_{3,i}}^{\frac{1}{q_2}} \\
e^*_{4,i} &= -\operatorname{sign}(\hat{u}_{1,i})\xi_{2,i}^{q_3} F_{2,i}(\cdot), & \xi_{3,i} &= e_{4,i}^{\frac{1}{q_3}} - {e^*_{4,i}}^{\frac{1}{q_3}} \\
&\vdots & &\vdots \\
e^*_{k+1,i} &= -\operatorname{sign}(\hat{u}_{1,i})\xi_{k-1,i}^{q_k} F_{k-1,i}(\cdot), & \xi_{k,i} &= e_{k+1,i}^{\frac{1}{q_k}} - {e^*_{k+1,i}}^{\frac{1}{q_k}}
\end{aligned}
$$

with $F_{1,i}(e_{2,i}) > 0$, $F_{2,i}(e_{2,i}, e_{3,i}) > 0$, …, $F_{k-1,i}(e_{2,i}, \ldots, e_{k,i}) > 0$ being C^1 functions, and a Lyapunov function $V_{k,i} = V_{2,i} + \int_{e^*_{3,i}}^{e_{3,i}} (s^{1/q_2} - {e^*_{3,i}}^{1/q_2})^{2-q_2}\, \mathrm{d}s + \cdots + \int_{e^*_{k,i}}^{e_{k,i}} (s^{1/q_{k-1}} - {e^*_{k,i}}^{1/q_{k-1}})^{2-q_{k-1}}\, \mathrm{d}s$ such that

$$
\begin{aligned}
\dot{V}_{k,i} \leq & -(m-k+1)|\hat{u}_{1,i}|(\xi_{1,i}^{q_2+1} + \xi_{2,i}^{q_2+1} + \cdots + \xi_{k-1,i}^{q_2+1}) \\
& - r|\hat{u}_{1,i}|(\xi_{1,i}^{q_2+\kappa+1} + \xi_{2,i}^{q_2+\kappa+1} + \cdots + \xi_{k-1,i}^{q_2+\kappa+1}) \\
& + \xi_{k-1,i}^{2-q_{k-1}} \hat{u}_{1,i}(e_{k+1,i} - e^*_{k+1,i}).
\end{aligned} \tag{6.25}
$$

In what follows, we will prove that (6.25) also holds at Step k. Construct a Lyapunov function as

$$
V_{k+1,i} = V_{k,i} + \int_{e^*_{k+1,i}}^{e_{k+1,i}} (s^{\frac{1}{q_k}} - {e^*_{k+1,i}}^{\frac{1}{q_k}})^{2-q_k}\, \mathrm{d}s. \tag{6.26}
$$

From (6.13) and (6.25), the time derivative of $V_{k+1,i}$ is

$$
\begin{aligned}
\dot{V}_{k+1,i} \leq & -(m-k+1)|\hat{u}_{1,i}|(\xi_{1,i}^{q_2+1} + \xi_{2,i}^{q_2+1} + \cdots + \xi_{k-1,i}^{q_2+1}) \\
& - r|\hat{u}_{1,i}|(\xi_{1,i}^{q_2+\kappa+1} + \xi_{2,i}^{q_2+\kappa+1} + \cdots + \xi_{k-1,i}^{q_2+\kappa+1}) \\
& + \xi_{k-1,i}^{2-q_{k-1}} \hat{u}_{1,i}(e_{k+1,i} - e^*_{k+1,i}) + \xi_{k,i}^{2-q_k} \hat{u}_{1,i} e_{k+2,i} \\
& - (2-q_k)\frac{\mathrm{d}{e^*_{k+1,i}}^{\frac{1}{q_k}}}{\mathrm{d}t} \int_{e^*_{k+1,i}}^{e_{k+1,i}} (s^{\frac{1}{q_k}} - {e^*_{k+1,i}}^{\frac{1}{q_k}})^{1-q_k}\, \mathrm{d}s.
\end{aligned} \tag{6.27}
$$

For the third term in (6.27), one gets

$$
\begin{aligned}
\xi_{k-1,i}^{2-q_{k-1}}\hat{u}_{1,i}(e_{k+1,i}-e_{k+1,i}^*) &\le |\hat{u}_{1,i}||\xi_{k-1,i}|^{2-q_{k-1}}\big|(e_{k+1,i}^{\frac{1}{q_k}})^{q_k}-(e_{k+1,i}^{*\,\frac{1}{q_k}})^{q_k}\big| \\
&\le 2^{1-q_k}|\hat{u}_{1,i}||\xi_{k-1,i}|^{2-q_{k-1}}|\xi_{k,i}|^{q_k} \\
&\le |\hat{u}_{1,i}|(\frac{1}{2}\xi_{k-1,i}^{q_2+1}+g_{k-1}\xi_{k,i}^{q_2+1})
\end{aligned}
\tag{6.28}
$$

for a positive constant g_{k-1}. We then provide the following fact.

Fact 6.2.1 There is a C^1 function $\tilde{F}_{k-1,i}(e_{2,i},\ldots,e_{k,i})>0$ such that $\big|\frac{\mathrm{d}e_{k+1,i}^{*\,1/q_k}}{\mathrm{d}t}\big| \le |\hat{u}_{1,i}|\big(|\xi_{1,i}|^{q_2}+|\xi_{2,i}|^{q_2}+\cdots+|\xi_{k,i}|^{q_2}\big)\tilde{F}_{k-1,i}(\cdot)$. □

More details about Fact 6.2.1 are provided in Appendix 6.5. For the last term in (6.27), one obtains

$$
\begin{aligned}
&\left|(2-q_k)\frac{\mathrm{d}e_{k+1,i}^{*\,\frac{1}{q_k}}}{\mathrm{d}t}\int_{e_{k+1,i}^*}^{e_{k+1,i}}(s^{\frac{1}{q_k}}-e_{k+1,i}^{*\,\frac{1}{q_k}})^{1-q_k}\,\mathrm{d}s\right| \\
&\le (2-q_k)|\hat{u}_{1,i}|\big(|\xi_{1,i}|^{q_2}+\cdots+|\xi_{k,i}|^{q_2}\big)\tilde{F}_{k-1}(\cdot)|e_{k+1,i}-e_{k+1,i}^*||\xi_{k,i}|^{1-q_k} \\
&\le 2^{1-q_k}(2-q_k)|\hat{u}_{1,i}||\xi_{k,i}|\big(|\xi_{1,i}|^{q_2}+\cdots+|\xi_{k,i}|^{q_2}\big)\tilde{F}_{k-1,i}(\cdot) \\
&\le |\hat{u}_{1,i}|\big(\frac{1}{2}(\xi_{1,i}^{q_2+1}+\xi_{2,i}^{q_2+1}+\cdots+\xi_{k-1,i}^{q_2+1})+\xi_{k,i}^{q_2+1}\hat{F}_{k-1,i}(\cdot)\big)
\end{aligned}
\tag{6.29}
$$

with $\hat{F}_{k-1,i}(e_{2,i},\ldots,e_{k,i})>0$ being a C^1 function. Combining (6.27), (6.28), and (6.29), one gets

$$
\begin{aligned}
\dot{V}_{k+1,i} \le &-(m-k)|\hat{u}_{1,i}|(\xi_{1,i}^{q_2+1}+\xi_{2,i}^{q_2+1}+\cdots+\xi_{k-1,i}^{q_2+1}) \\
&-r|\hat{u}_{1,i}|(\xi_{1,i}^{q_2+\kappa+1}+\xi_{2,i}^{q_2+\kappa+1}+\cdots+\xi_{k-1,i}^{q_2+\kappa+1}) \\
&+\xi_{k,i}^{2-q_k}\hat{u}_{1,i}(e_{k+2,i}-e_{k+2,i}^*)+\xi_{k,i}^{2-q_k}\hat{u}_{1,i}e_{k+2,i}^* \\
&+|\hat{u}_{1,i}|\xi_{k,i}^{q_2+1}\big(g_{k-1}+\hat{F}_{k-1,i}(e_{2,i},\ldots,e_{k,i})\big).
\end{aligned}
\tag{6.30}
$$

Design a virtual protocol as

$$
\begin{aligned}
e_{k+2,i}^* &= -\operatorname{sign}(\hat{u}_{1,i})\xi_{k,i}^{q_{k+1}}\big(m-k+r\xi_{k,i}^{\kappa}+g_{k-1}+\hat{F}_{k-1,i}(e_{2,i},\ldots,e_{k,i})\big) \\
&\triangleq -\operatorname{sign}(\hat{u}_{1,i})\xi_{k,i}^{q_{k+1}}F_{k,i}(e_{2,i},\ldots,e_{k+1,i})
\end{aligned}
\tag{6.31}
$$

with $F_{k,i}(e_{2,i},\ldots,e_{k+1,i})>0$ being a C^1 function. Substituting (6.31) into (6.30) yields

$$
\begin{aligned}
\dot{V}_{k+1,i} \le &-(m-k)|\hat{u}_{1,i}|(\xi_{1,i}^{q_2+1}+\xi_{2,i}^{q_2+1}+\cdots+\xi_{k,i}^{q_2+1}) \\
&-r|\hat{u}_{1,i}|(\xi_{1,i}^{q_2+\kappa+1}+\xi_{2,i}^{q_2+\kappa+1}+\cdots+\xi_{k,i}^{q_2+\kappa+1}) \\
&+\xi_{k,i}^{2-q_k}\hat{u}_{1,i}(e_{k+2,i}-e_{k+2,i}^*).
\end{aligned}
\tag{6.32}
$$

Based on the above inductive argument, at the last step for (6.13), there exist C^1 functions $F_{1,i}(e_{2,i}) > 0$, $F_{2,i}(e_{2,i}, e_{3,i}) > 0$, ..., $F_{m-2,i}(e_{2,i}, \ldots, e_{m-1,i}) > 0$, $\hat{F}_{m-2,i}(e_{2,i}, \ldots, e_{m-1,i}) > 0$, and a constant $g_{m-2} > 0$ such that the following inequality holds:

$$\begin{aligned}\dot{V}_{m,i} \leq &- |\hat{u}_{1,i}|(\xi_{1,i}^{q_2+1} + \xi_{2,i}^{q_2+1} + \cdots + \xi_{m-2,i}^{q_2+1}) + \xi_{m-1,i}^{2-q_{m-1}}(u_{2,i} - \hat{u}_{2,i}) \\ &- r|\hat{u}_{1,i}|(\xi_{1,i}^{q_2+\kappa+1} + \xi_{2,i}^{q_2+\kappa+1} + \cdots + \xi_{m-2,i}^{q_2+\kappa+1}) \\ &+ |\hat{u}_{1,i}|\xi_{m-1,i}^{q_2+1}\big(g_{m-2} + \hat{F}_{m-2,i}(e_{2,i}, \ldots, e_{m-1,i})\big),\end{aligned} \tag{6.33}$$

where

$$V_{m,i} = V_{2,i} + \int_{e_{3,i}^*}^{e_{3,i}} \left(s^{\frac{1}{q_2}} - e_{3,i}^{*\frac{1}{q_2}}\right)^{2-q_2} \mathrm{d}s + \cdots + \int_{e_{m,i}^*}^{e_{m,i}} \left(s^{\frac{1}{q_{m-1}}} - e_{m,i}^{*\frac{1}{q_{m-1}}}\right)^{2-q_{m-1}} \mathrm{d}s.$$

Then, by using protocol (6.15), (6.33) can be rewritten as

$$\begin{aligned}\dot{V}_{m,i} \leq &- |\hat{u}_{1,i}|(\xi_{1,i}^{q_2+1} + \xi_{2,i}^{q_2+1} + \cdots + \xi_{m-1,i}^{q_2+1}) \\ &- r|\hat{u}_{1,i}|(\xi_{1,i}^{q_2+\kappa+1} + \xi_{2,i}^{q_2+\kappa+1} + \cdots + \xi_{m-1,i}^{q_2+\kappa+1}).\end{aligned} \tag{6.34}$$

By using Lemma 6.1, for $z = 3, 4, \ldots, m$, one obtains

$$\begin{aligned}\int_{e_{z,i}^*}^{e_{z,i}} (s^{\frac{1}{q_{z-1}}} - e_{z,i}^{*\frac{1}{q_{z-1}}})^{2-q_{z-1}} \mathrm{d}s &\leq |e_{z,i} - e_{z,i}^*||\xi_{z-1,i}|^{2-q_{z-1}} \\ &\leq 2^{1-q_{z-1}}|\xi_{z-1,i}|^{q_{z-1}}|\xi_{z-1,i}|^{2-q_{z-1}} \leq 2\xi_{z-1,i}^2.\end{aligned} \tag{6.35}$$

Then, one concludes that $V_{m,i} \leq 2(\xi_{1,i}^2 + \xi_{2,i}^2 + \cdots + \xi_{m-1,i}^2)$. Since $0 < \frac{q_2+1}{2} < 1$, by using Lemma 3.3, one gets

$$V_{m,i}^{\frac{q_2+1}{2}} \leq 2^{\frac{q_2+1}{2}}(\xi_{1,i}^{q_2+1} + \xi_{2,i}^{q_2+1} + \cdots + \xi_{m-1,i}^{q_2+1}). \tag{6.36}$$

Since $\frac{q_2+\kappa+1}{2} > 1$, by using Lemma 3.3, one obtains

$$\begin{aligned}&\xi_{1,i}^{q_2+\kappa+1} + \xi_{2,i}^{q_2+\kappa+1} + \cdots + \xi_{m-1,i}^{q_2+\kappa+1} \\ &= (\xi_{1,i}^2)^{\frac{q_2+\kappa+1}{2}} + (\xi_{2,i}^2)^{\frac{q_2+\kappa+1}{2}} + \cdots + (\xi_{m-1,i}^2)^{\frac{q_2+\kappa+1}{2}} \\ &\geq (m-1)^{\frac{1-q_2-\kappa}{2}}(\xi_{1,i}^2 + \xi_{2,i}^2 + \cdots + \xi_{m-1,i}^2)^{\frac{q_2+\kappa+1}{2}} \\ &\geq 2^{-\frac{q_2+\kappa+1}{2}}(m-1)^{\frac{1-q_2-\kappa}{2}}V_{m,i}^{\frac{q_2+\kappa+1}{2}}.\end{aligned} \tag{6.37}$$

Combining (6.34), (6.36), and (6.37) yields

$$\dot{V}_{m,i} \leq -|\hat{u}_{1,i}|2^{-\frac{q_2+1}{2}}V_{m,i}^{\frac{q_2+1}{2}} - r|\hat{u}_{1,i}|2^{-\frac{q_2+\kappa+1}{2}}(m-1)^{\frac{1-q_2-\kappa}{2}}V_{m,i}^{\frac{q_2+\kappa+1}{2}}.$$

Since $\kappa = 2 - 2q_2$ holds,

$$\dot{V}_{m,i} \leq -|\hat{u}_{1,i}|2^{-\frac{q_2+1}{2}} V_{m,i}^{1-\frac{1-q_2}{2}} - r|\hat{u}_{1,i}|2^{-\frac{3-q_2}{2}}(m-1)^{\frac{q_2-1}{2}} V_{m,i}^{1+\frac{1-q_2}{2}}.$$

Then, by using Theorem 2.4, $V_{m,i}$ converges to zero in a finite time bounded by $T_{\max 2}$, which implies that (6.13) is fixed-time stable. Therefore, consensus tracking is realized within a fixed-time satisfying (6.16). □

Remark 6.2 We do not prove that protocols (6.12) and (6.15) together with observer (6.4) can guarantee the boundedness of tracking errors on the interval $[0, T_{\max 1} + T_2]$ since the nonlinear terms are too complex. Inspired by [4], for $t \in [0, T_{\max 1} + T_2]$, one can design other forms of $u_{1,i}(t)$ and $u_{2,i}(t)$ such that $e_{k,i}(t)$ is bounded. It is worth mentioning that, unlike the switching instant in [4] that depends on initial states, $T_{\max 1}$ and T_2 derived in this chapter are independent on initial states, which facilitates the implementation for protocols.

6.3 A Numerical Example

The results in Sect. 6.2 are applied to wheeled mobile robots. Without loss of generality, it is assumed that there are four followers and one leader, i.e., $N = 4$. An undirected graph shown in Fig. 6.1 (Left) is used as the interaction topology.

The model of the ith, $\forall i \in \{0, 1, 2, 3, 4\}$, wheeled mobile robot, shown in Fig. 6.2 (Right), is described by [5]

$$\dot{x}_i = v_i \cos(\theta_i), \quad \dot{y}_i = v_i \sin(\theta_i), \quad \dot{\theta}_i = \omega_i, \tag{6.38}$$

where $(x_i, y_i) \in \mathbb{R}^2$ is the Cartesian position of the ith robot center, θ_i is the orientation, and v_i and ω_i are the linear velocity and the angular velocity, respectively.

Using the coordinate transformation [1] of $x_{1,i} = -\theta_i$, $x_{2,i} = -x_i \sin(\theta_i) + y_i \cos(\theta_i)$, $x_{3,i} = x_i \cos(\theta_i) + y_i \sin(\theta_i)$, $u_{1,i} = -\omega_i$, $u_{2,i} = v_i + x_{2,i}\omega_i$, $\forall i \in \{0, 1, 2, 3, 4\}$, the system (6.38) is transformed to

$$\dot{x}_{1,i} = u_{1,i}, \quad \dot{x}_{2,i} = x_{3,i}u_{1,i}, \quad \dot{x}_{3,i} = u_{2,i}. \tag{6.39}$$

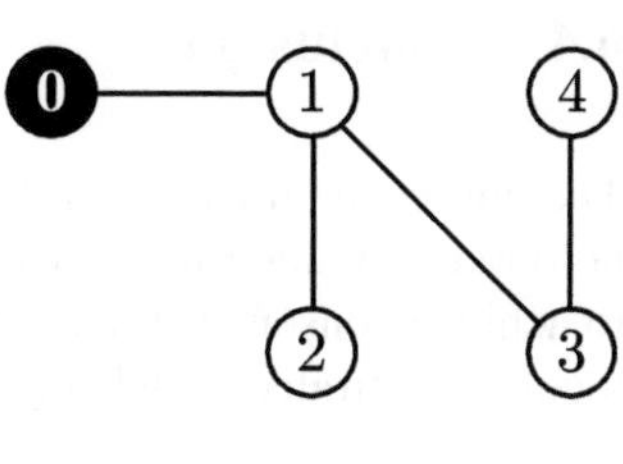

Fig. 6.1 An interaction graph

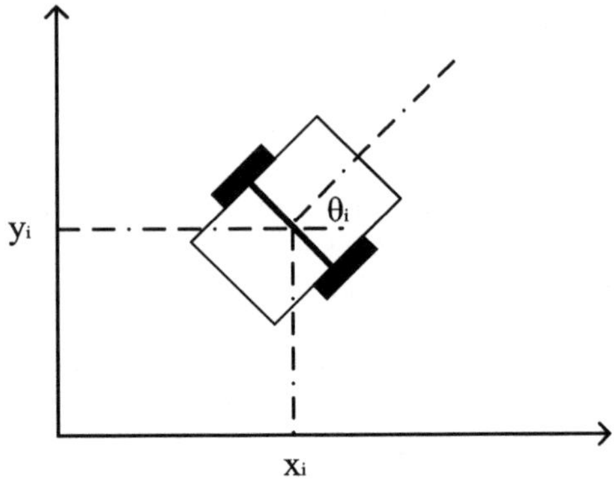

Fig. 6.2 A wheeled mobile robot i

For system (6.39), an observer (6.4) with $m = 3$ is implemented together with $\alpha_0 = \beta_0 = \gamma_0 = \chi_0 = 1, a = 2$, and $b = 5$. Set $r = 0.3, \sigma = -2/5, \kappa = 4/5$. The input of leader is selected as $u_{1,0} = 0.1\cos(2\pi t) + 1$, $u_{2,0} = 0.3\sin(2\pi t) + 0.5$, from which it can be obtained that $c_3 = 0.9, \rho_1 = 1.1$. Besides, $\zeta_1 = \zeta_2 = 2.5$ are selected. Then, protocols in the form of (6.12) and (6.15), respectively, are implemented as

$$\begin{cases} u_{1,i} = \hat{u}_{1,i} - 2^{-1+\frac{a}{2b}}\alpha_0 e_{1,i}^{1-\frac{a}{b}} - 2^{-1-\frac{a}{2b}}\beta_0 e_{1,i}^{1+\frac{a}{b}}; \\ u_{2,i} = \hat{u}_{2,i} - |\hat{u}_{1,i}|\xi_{2,i}^{q_3}\left(1 + r\xi_{2,i}^{\kappa} + g_1 + \hat{F}_1(e_{2,i})\right) \end{cases} \tag{6.40}$$

with $g_1 = 1.140$, and $\hat{F}_{1,i}(e_{2,i}) = 1.85\tilde{F}_{1,i}(e_{2,i})(0.34\tilde{F}_{1,i}^{3/5}(e_{2,i}) + 1)$, where $\tilde{F}_{1,i}(e_{2,i}) = (3 + 0.3e_{2,i}^{4/5})[(2 + 0.3e_{2,i}^{4/5})^{5/3} + 0.4e_{2,i}^{4/5}(2 + 0.3e_{2,i}^{4/5})^{2/3}]$. Let the initial states of four followers as $x(0) = [(0.3, -0.2, 0.2), (-0.1, 0.2, -0.1), (-0.3, -0.3, -0.2), (0.3, -0.1, 0.1)]^T$, and let the initial states of the leader as $x_0(0) = [(1, 1, 1)]^T$. It can be calculated that $T_{\max1} = 23.56$ s and $T_{\max2} = 34.15$ s. Therefore, the settling time is upper bounded by $T_{\max1} + T_{\max2} + T_2 = 65.56$ s by Theorem 6.2. Implementing the protocol (6.40) together with the observer (6.4) with $m = 3$, in Figs. 6.3, 6.4, and 6.5, the settling time is about $t = 4.75$ s, which is smaller than 65.56 s. Then, change the initial states of followers to $x(0) = [(2, 2, -2), (-3, -5, 5), (1, 1, 1), (-2, -1, -2)]^T$. The results are shown in Figs. 6.6, 6.7, and 6.8, where the settling time is about $t = 5.23$ s, which is still smaller than 65.56 s.

6.4 Conclusions

The consensus tracking problem is investigated for multi-agent systems with nonholonomic chained-form dynamics. An observer-based nonlinear protocol is designed to achieve consensus tracking in fixed time. The effectiveness of the proposed protocol is illustrated by wheeled mobile robots.

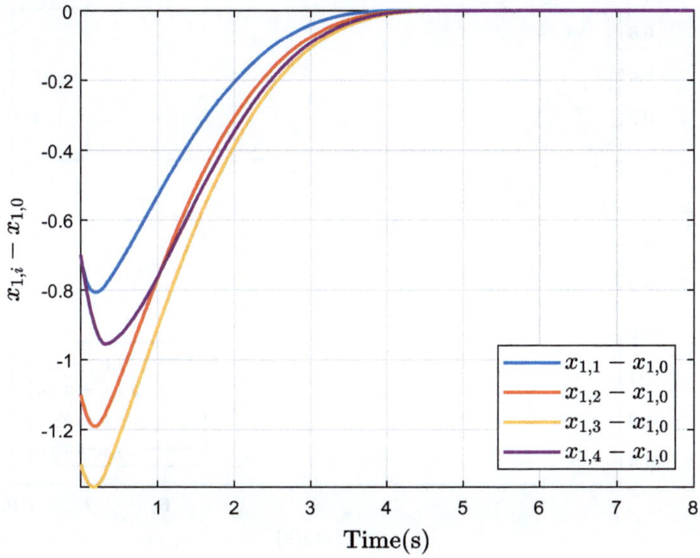

Fig. 6.3 $x_{1,i} - x_{1,0}$

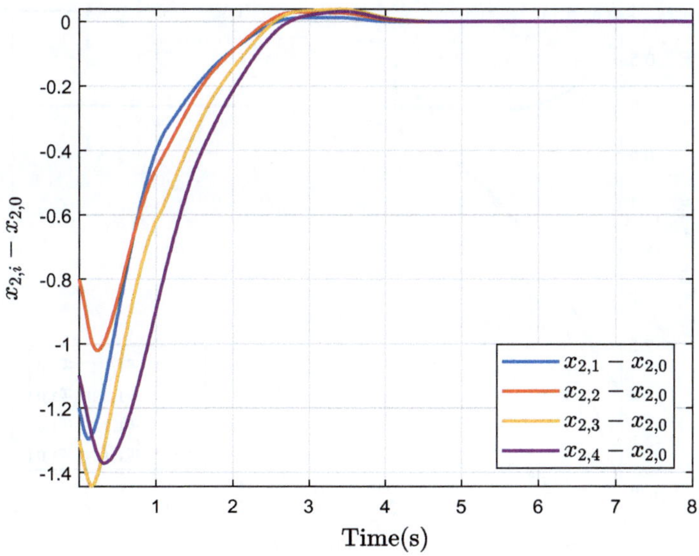

Fig. 6.4 $x_{2,i} - x_{2,0}$

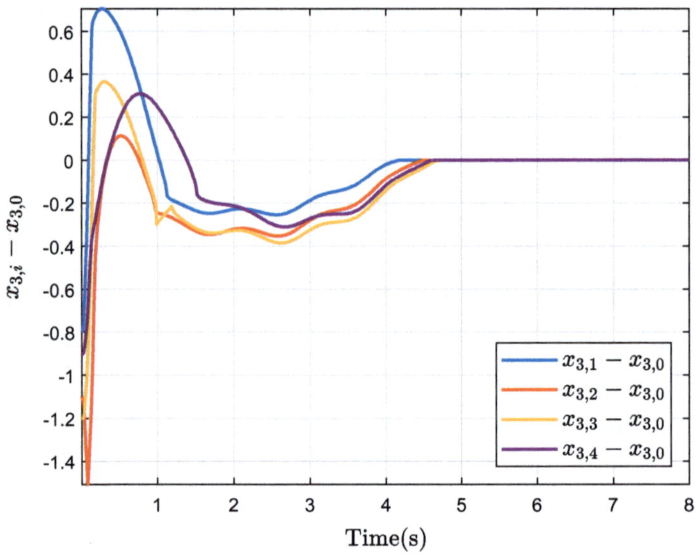

Fig. 6.5 $x_{3,i} - x_{3,0}$

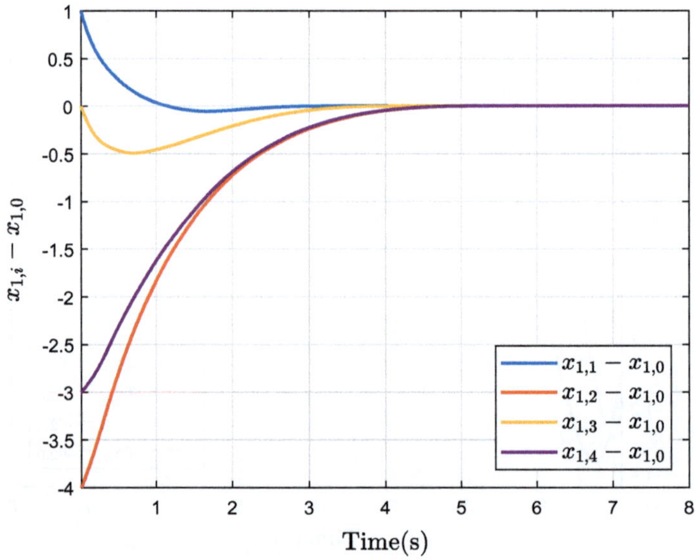

Fig. 6.6 $x_{1,i} - x_{1,0}$

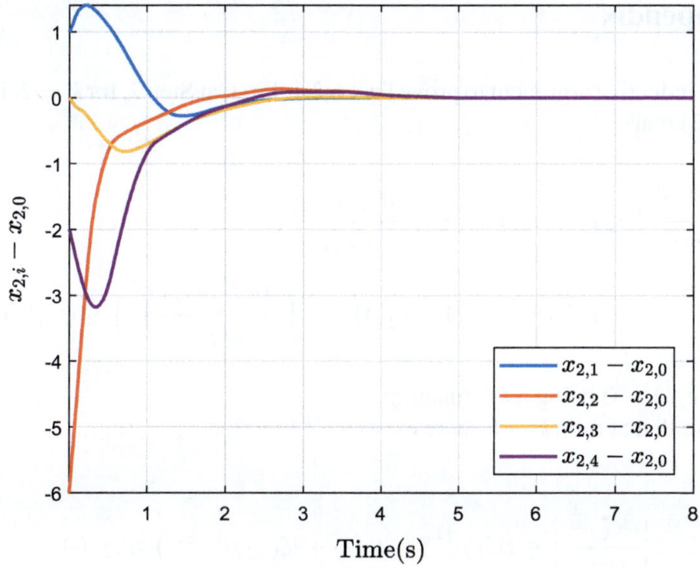

Fig. 6.7 $x_{2,i} - x_{2,0}$

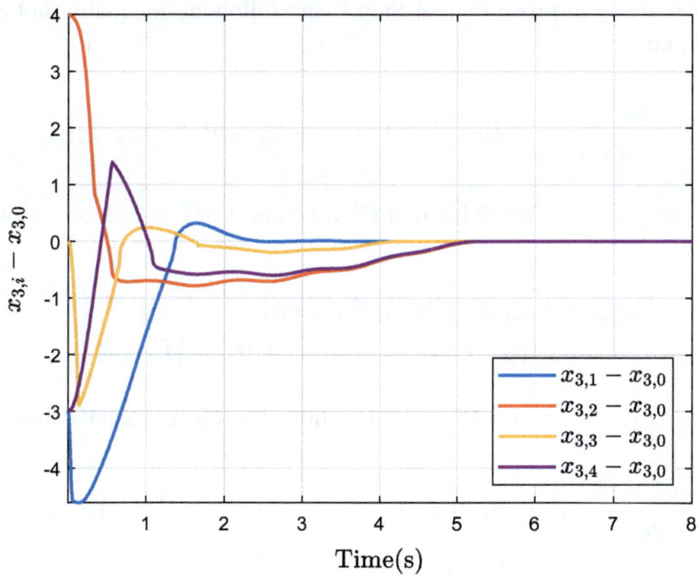

Fig. 6.8 $x_{3,i} - x_{3,0}$

6.5 Appendix

We use an inductive argument to prove Fact 6.2.1. First, at Step 2, for $k=2$, it follows from (6.22) that

$$\begin{aligned}\left|\frac{\partial e_{3,i}^{*\frac{1}{q_2}}}{\partial e_{2,i}}\right| &\le F_{1,i}^{\frac{1}{q_2}}(e_{2,i})+|e_{2,i}|\left|\frac{\mathrm{d}F_{1,i}^{\frac{1}{q_2}}(e_{2,i})}{\mathrm{d}e_{2,i}}\right|\\ &\le F_{1,i}^{\frac{1}{q_2}}(e_{2,i})+(1+e_{2,i}^2)\left(1+\left(\frac{\mathrm{d}F_{1,i}^{\frac{1}{q_2}}(e_{2,i})}{\mathrm{d}e_{2,i}}\right)^2\right)\le \varphi_{1,i}(e_{2,i})\end{aligned}$$

with $\varphi_{1,i}(e_{2,i})>0$ being a C^1 function.

Suppose that at Step $k-1$, there exists a C^1 function $\varphi_{k-2,i}(e_{2,i},\ldots,e_{k-1,i})>0$ such that

$$\left|\frac{\partial e_{k,i}^{*\frac{1}{q_k}}}{\partial e_{l,i}}\right|\le \big(|\xi_{1,i}|^{1-q_{l-1}}+\cdots+|\xi_{k-2,i}|^{1-q_{l-1}}\big)\varphi_{k-2,i}(\cdot)$$

holds for $l=2,3,\ldots,k-1$. Following the same line in the proof of Proposition 4 in [6], it is ready to prove that, at Step k, the following inequality holds for $l=2,3,\ldots,k$, i.e.,

$$\left|\frac{\partial e_{k+1,i}^{*\frac{1}{q_k}}}{\partial e_{l,i}}\right|\le \big(|\xi_{1,i}|^{1-q_{l-1}}+\cdots+|\xi_{k-1,i}|^{1-q_{l-1}}\big)\varphi_{k-1,i}(\cdot)$$

with $\varphi_{k-1,i}(e_{2,i},\ldots,e_{k,i})>0$ being a C^1 function. For $l=2,3,\ldots,k$, we obtain that

$$\begin{aligned}|e_{l+1,i}| &\le |\xi_{l,i}|^{q_l}+|\xi_{l-1,i}|^{q_l}F_{l-1,i}(e_{2,i},\ldots,e_{l,i})\\ &\le \big(|\xi_{1,i}|^{q_l}+|\xi_{2,i}|^{q_l}+\cdots+|\xi_{k,i}|^{q_l}\big)\bar{F}_{l-1,i}(\cdot),\end{aligned}$$

where $\bar{F}_{l-1,i}(e_{2,i},\ldots,e_{l,i})>0$ is a C^1 function. Therefore, we have that

$$\begin{aligned}\left|\frac{\partial e_{k+1,i}^{*\frac{1}{q_k}}}{\partial e_{l,i}}\dot{e}_{l,i}\right| &\le |\hat{u}_{1,i}|\big(|\xi_{1,i}|^{1-q_{l-1}}+\cdots+|\xi_{k-1,i}|^{1-q_{l-1}}\big)\varphi_{k-1,i}(\cdot)\\ &\quad\cdot\big(|\xi_{1,i}|^{q_l}+\cdots+|\xi_{k,i}|^{q_l}\big)\bar{F}_{l-1,i}(\cdot)\\ &\le |\hat{u}_{1,i}|\big(|\xi_{1,i}|^{q_2}+\cdots+|\xi_{k,i}|^{q_2}\big)\breve{F}_{l-1,i}(\cdot),\end{aligned}$$

where $\breve{F}_{l-1,i}(e_{2,i},\ldots,e_{k,i})>0$ is a C^1 function. Then, we obtain that

$$
\begin{aligned}
\left|\frac{\mathrm{d}e_{k+1,i}^{*\frac{1}{q_k}}}{\mathrm{d}t}\right| &= \sum_{l=2}^{k}\left|\frac{\partial e_{k+1,i}^{*\frac{1}{q_k}}}{\partial e_{l,i}}\dot{e}_{l,i}\right| \\
&\le |\hat{u}_{1,i}|\big(|\xi_{1,i}|^{q_2}+\cdots+|\xi_{k,i}|^{q_2}\big)\sum_{l=2}^{k}\breve{F}_{l-1,i}(\cdot) \\
&\le |\hat{u}_{1,i}|\big(|\xi_{1,i}|^{q_2}+\cdots+|\xi_{k,i}|^{q_2}\big)\tilde{F}_{k-1,i}(\cdot),
\end{aligned}
$$

which completes the proof.

6.6 Notes

The adding a power integrator technique is used in [7], where protocols are designed for the first time to stabilize an uncertain nonholonomic system in fixed time. Based on the work in [8], in this chapter, we are dealing with a consensus tracking problem for multi-agent systems with N agents. First, observers are proposed under undirected communication graphs. Some Lyapunov functions are constructed and corresponding observer gains are derived to ensure that the leader's state and the leader's input are accurately estimated by each follower. Second, an observer-based protocol is proposed to achieve consensus tracking in fixed time. In [9], a fixed-time observer is designed for high-order multi-agent systems under undirected communication graphs. An explicit bound for settling time is not provided, while in this chapter, we provided such a bound in Theorem 6.2. Unlike the protocols proposed in this chapter, some time-varying protocols are proposed in [10] to achieve coordinated tracking for high-order multi-agent systems. The time-varying protocols [11] are able to overcome non-vanish uncertainties. It would be an interesting research direction to solve a consensus tracking problem for multi-agent systems with nonholonomic chained-form dynamics by using time-varying protocols.

References

1. W. Dong, J.A. Farrell, Cooperative control of multiple nonholonomic mobile agents. IEEE Trans. Autom. Control **53**(6), 1434–1448 (2008)
2. C. Qian, W. Lin, A continuous feedback approach to global strong stabilization of nonlinear systems. IEEE Trans. Autom. Control **46**(7), 1061–1079 (2001)
3. B. Ning, J. Jin, J. Zheng, Z. Man, Finite-time and fixed-time leader-following consensus for multi-agent systems with discontinuous inherent dynamics. Int. J. Control **91**(6), 1259–1270 (2018)
4. S. Li, H. Du, X. Lin, Finite-time consensus algorithm for multi-agent systems with double-integrator dynamics. Automatica **47**(8), 1706–1712 (2011)
5. W. Ren, E.M. Atkins, Distributed multi-vehicle coordinated control via local information exchange. Int. J. Robust Nonlinear Control **17**, 1002–1033 (2007)

6. X. Huang, W. Lin, B. Yang, Global finite-time stabilization of a class of uncertain nonlinear systems. Automatica **41**(5), 881–888 (2005)
7. Z. Zhang, Y. Wu, Fixed-time regulation control of uncertain nonholonomic systems and its applications. Int. J. Control **90**(7), 1327–1344 (2017)
8. B. Ning, Q.-L. Han, Prescribed finite-time consensus tracking for multi-agent systems with nonholonomic chained-form dynamics. IEEE Trans. Autom. Control **64**(4), 1686–1693 (2019)
9. Z. Zuo, B. Tian, M. Defoort, Z. Ding, Fixed-time consensus tracking for multi-agent systems with high-order integrator dynamics. IEEE Trans. Autom. Control **63**(2), 563–570 (2018)
10. Y. Wang, Y. Song, Leader-following control of high-order multi-agent systems under directed graphs: pre-specified finite time approach. Automatica **87**, 113–120 (2018)
11. Y. Song, Y. Wang, J. Holloway, M. Krstic, Time-varying feedback for regulation of normal-form nonlinear systems in prescribed finite time. Automatica **83**, 243–251 (2017)

Chapter 7
Distributed Optimization: An Edge-Based Fixed-Time Consensus Approach

This chapter deals with the problem of distributed optimization for multi-agent systems by using an edge-based fixed-time consensus approach. In the case of time-invariant cost functions, a distributed protocol is proposed to achieve the state agreement in a fixed time while the sum of local convex functions known to individual agents are minimized. In the case of time-varying cost functions, based on the distributed protocol in the case of time-invariant cost functions, a distributed protocol is provided by taking the Hessian matrix into account. In both cases, stability conditions are derived to ensure that the distributed optimization problem is solved under both fixed and switching communication topologies. A distinctive feature of the results in this chapter is that an upper bound of settling time for consensus can be estimated without dependence on initial states of agents, and thus can be made arbitrarily small through adjusting system parameters. Therefore, the results in this chapter can be applicable in an unknown environment such as drone rendezvous within a required time for military purpose while optimizing local objectives. Case studies of a power output agreement for battery packages are provided to demonstrate the effectiveness of the theoretical results.

7.1 Problem Formulation

Consider a multi-agent system, which consists of N agents labeled with index $1, 2, \ldots, N$, and each agent evolves according to the dynamics of

$$\dot{x}_i(t) = u_i(t), \quad i \in \Gamma, \tag{7.1}$$

where $x_i(t) \in \mathbb{R}^m$ is the ith agent's state, and $u_i(t) \in \mathbb{R}^m$ is the protocol.

Z. Zuo et al., *Fixed-Time Cooperative Control of Multi-Agent Systems*,
https://doi.org/10.1007/978-3-030-20279-8_7

The objective is to design $u_i(t)$ by using only local interaction and information such that all agents cooperatively reach the optimal state x^* that solves a time-invariant convex optimization problem

$$\text{minimize} \sum_{i=1}^{N} f_i(x), \quad \text{subject to} \quad x \in \mathbb{R}^m, \tag{7.2}$$

or a time-varying convex optimization problem

$$\text{minimize} \sum_{i=1}^{N} f_i(x,t), \quad \text{subject to} \quad x \in \mathbb{R}^m, \tag{7.3}$$

where both $f_i(x) : \mathbb{R}^m \to \mathbb{R}$ and $f_i(x,t) : \mathbb{R}^m \times \mathbb{R}^+ \to \mathbb{R}$ denote the local objective function of agent i, which is only known to agent i itself.

We introduce the following definition and lemmas that will be used in Sects. 7.2 and 7.3.

Definition 7.1 For the system (7.1), it is said to achieve fixed-time consensus if there exists a suitable protocol $u_i(t)$ such that

$$\begin{cases} \lim\limits_{t \to T} \|x_i(t) - x_j(t)\| = 0; \\ \|x_i(t) - x_j(t)\| = 0, \quad \forall t \geq T \end{cases}$$

for all $i, j \in \Gamma$, where T is upper bounded by $T_{\max}$ that is independent of initial states.

Lemma 7.1 ([1]) *Let $f_0(x) : \mathbb{R}^m \to \mathbb{R}$ be a continuously differentiable convex function, then $f_0(x)$ is minimized if and only if $\nabla f_0(x) = 0$.*

7.2 Distributed Optimization with Time-Invariant Cost Functions

In this section, we consider a distributed optimization problem with a class of time-invariant cost functions, namely (7.2). The following assumption is used in this section.

Assumption 7.2.1 For the cost function $f_i(x)$, it is differentiable with respect to x, and the optimal solution set of each $f_i(x)$ is nonempty. In particular, $\nabla f_i(x(t)) = \rho x(t) + \varphi_i(x(t))$ holds $\forall x(t) \in \mathbb{R}^m$, where $\rho \geq 0$ and $\|\varphi_i\|_2 \leq g_0$ (g_0 is a known positive constant). □

Note that (7.2) is equivalent to the problem that all agents reach a consensus while minimizing the overall cost function $\sum_{i=1}^{N} f_i(x_i)$, i.e.,

$$\text{minimize} \quad \sum_{i=1}^{N} f_i(x_i), \quad \text{subject to} \quad x_i = x_j \in \mathbb{R}^m. \tag{7.4}$$

We propose a distributed edge-based protocol for the ith agent, i.e.,

$$\begin{aligned} u_i = & -\alpha_1 \sum_{j=1}^{N} a_{ij}(x_i - x_j)^{1-\frac{a}{b}} - \beta_1 \sum_{j=1}^{N} a_{ij}(x_i - x_j)^{1+\frac{a}{b}} \\ & - \gamma_1 \sum_{j=1}^{N} a_{ij}\text{sign}(x_i - x_j) - k_1 \nabla f_i(x_i), \end{aligned} \tag{7.5}$$

where k_1 is a positive constant, a and b are a positive even integer and a positive odd integer, respectively, satisfying $a < b$. In addition, α_1, β_1, and γ_1 are feedback gains.

Remark 7.1 To solve consensus problems, there are basically two types of protocols: one is a node-based protocol, which, for example, takes the following form:

$$\hat{u}_i = -\alpha_1 \left(\sum_{j=1}^{N} a_{ij}(x_i - x_j) \right)^{1-\frac{a}{b}} - \beta_1 \left(\sum_{j=1}^{N} a_{ij}(x_i - x_j) \right)^{1+\frac{a}{b}};$$

the other is an edge-based protocol, which is of the form

$$\hat{u}_i = -\alpha_1 \sum_{j=1}^{N} a_{ij}(x_i - x_j)^{1-\frac{a}{b}} - \beta_1 \sum_{j=1}^{N} a_{ij}(x_i - x_j)^{1+\frac{a}{b}}.$$

In protocol (7.5), the first three terms aim to achieve the fixed-time consensus, while the last term is for the optimization purpose. Note that the first three terms are all in the edge-based form. Compared to the node-based protocol used in [2, 3], the edge-based protocol developed in this chapter guarantees certain symmetry without affecting the optimization process. The symmetric property enables the establishment of stability conditions for networked agents with both fixed and switching topologies.

Remark 7.2 Under Assumption 7.2.1, the requirement of $\nabla f_i(x(t)) = \rho x(t) + \varphi_i(x(t))$ indicates that $f_i(x(t))$ belongs to a class of quadratic convex functions. In fact, this assumption is weaker than the one in [4], where the subgradient of $f_i(x(t))$ is assumed to be bounded. In other words, we are dealing with a more generalized class of objective functions.

In what follows, we first prove that agents will reach a consensus in a fixed time using (7.5). Then we prove that the optimization part for $\sum_{i=1}^{N} f_i(x)$ is solved accordingly. Let $\mathscr{L}_B$ and $\mathscr{L}_C$ denote the Laplacian matrix of graph $\mathscr{G}([a_{ij}^{2b/(2b-a)}])$ and $\mathscr{G}([a_{ij}^{2b/(2b+a)}])$, respectively.

Lemma 7.2 *For a multi-agent system (7.1), the communication topology* $\mathcal{G}$ *is connected and fixed, and Assumption 7.2.1 holds. If* $\gamma_1 \geq 2Nk_1 g_0/a_0$, *where* $a_0 = \min\{a_{ij} : a_{ij} \neq 0, \forall i, j \in \Gamma\}$, *then the protocol (7.5) enables the system to achieve fixed-time consensus. Moreover, the settling time* $T \leq T_1$, *where*

$$T_1 = \frac{2\pi b m^{\frac{a}{4b}} N^{\frac{a}{2b}}}{a\sqrt{\alpha_1\beta_1}(4\lambda_2(\mathcal{L}_B))^{\frac{1}{2}-\frac{a}{4b}}(4\lambda_2(\mathcal{L}_C))^{\frac{1}{2}+\frac{a}{4b}}}. \tag{7.6}$$

Proof Define $e_i = x_i - \frac{1}{N}\sum_{j=1}^N x_j$. Choose a Lyapunov function candidate as $V = \frac{1}{2}\sum_{i=1}^N e_i^T e_i$. Its time derivative is given by

$$\begin{aligned}
\dot{V} =& \sum_{i=1}^N e_i^T \dot{e}_i \\
=& -\alpha_1 \sum_{i=1}^N \sum_{j=1}^N a_{ij} e_i^T (x_i - x_j)^{1-\frac{a}{b}} - \frac{1}{N}\sum_{i=1}^N \sum_{j=1}^N e_i^T \dot{x}_j \\
& - \beta_1 \sum_{i=1}^N \sum_{j=1}^N a_{ij} e_i^T (x_i - x_j)^{1+\frac{a}{b}} - k_1 \sum_{i=1}^N e_i^T \varphi_i(x_i) \\
& - \gamma_1 \sum_{i=1}^N \sum_{j=1}^N a_{ij} e_i^T \operatorname{sign}(x_i - x_j) - k_1\rho \sum_{i=1}^N e_i^T x_i.
\end{aligned} \tag{7.7}$$

Note that, for the second term in (7.7), one gets

$$-\frac{1}{N}\sum_{i=1}^N e_i^T \left(\sum_{j=1}^N \dot{x}_j\right) = 0 \tag{7.8}$$

due to the fact that $\sum_{i=1}^N e_i^T = \mathbf{0}^T$. Then, for the last term in (7.7), one obtains

$$\begin{aligned}
-k_1\rho \sum_{i=1}^N e_i^T x_i =& -k_1\rho \sum_{i=1}^N e_i^T \left(e_i + \frac{1}{N}\sum_{j=1}^N x_j\right) \\
=& -k_1\rho \sum_{i=1}^N e_i^T e_i = -k_1\rho \sum_{i=1}^N \|e_i\|_2^2,
\end{aligned} \tag{7.9}$$

where the fact that $\sum_{i=1}^N e_i^T (\frac{1}{N}\sum_{j=1}^N x_j) = 0$ is used. Moreover, for the first term in (7.7),

$$
\begin{aligned}
& - \alpha_1 \sum_{i=1}^{N} \sum_{j=1}^{N} a_{ij} e_i^T (x_i - x_j)^{1-\frac{a}{b}} \\
= & - \alpha_1 \sum_{i=1}^{N} \sum_{j=1}^{N} a_{ij} e_i^T (e_i - e_j)^{1-\frac{a}{b}} \\
= & - \frac{\alpha_1}{2} \sum_{i=1}^{N} \sum_{j=1}^{N} a_{ij} e_i^T (e_i - e_j)^{1-\frac{a}{b}} - \frac{\alpha_1}{2} \sum_{j=1}^{N} \sum_{i=1}^{N} a_{ji} e_j^T (e_j - e_i)^{1-\frac{a}{b}} \\
= & - \frac{\alpha_1}{2} \sum_{i=1}^{N} \sum_{j=1}^{N} a_{ij} e_i^T (e_i - e_j)^{1-\frac{a}{b}} + \frac{\alpha_1}{2} \sum_{i=1}^{N} \sum_{j=1}^{N} a_{ij} e_j^T (e_i - e_j)^{1-\frac{a}{b}} \\
= & - \frac{\alpha_1}{2} \sum_{i=1}^{N} \sum_{j=1}^{N} a_{ij} \left\| e_i - e_j \right\|_{2-\frac{a}{b}}^{2-\frac{a}{b}}.
\end{aligned} \tag{7.10}
$$

Following the same line, one gets

$$
-\beta_1 \sum_{i=1}^{N} \sum_{j=1}^{N} a_{ij} e_i^T (x_i - x_j)^{1+\frac{a}{b}} = -\frac{\beta_1}{2} \sum_{i=1}^{N} \sum_{j=1}^{N} a_{ij} \left\| e_i - e_j \right\|_{2+\frac{a}{b}}^{2+\frac{a}{b}}. \tag{7.11}
$$

Then, for the fifth term in (7.7), one obtains

$$
\begin{aligned}
-\gamma_1 \sum_{i=1}^{N} \sum_{j=1}^{N} a_{ij} e_i^T \operatorname{sign}(x_i - x_j) &= -\frac{\gamma_1}{2} \sum_{i=1}^{N} \sum_{j=1}^{N} a_{ij} (e_i - e_j)^T \operatorname{sign}(e_i - e_j) \\
&= -\frac{\gamma_1}{2} \sum_{i=1}^{N} \sum_{j=1}^{N} a_{ij} \left\| e_i - e_j \right\|_1.
\end{aligned} \tag{7.12}
$$

Next, we will derive an upper bound for $\|e_i\|_2$. First, suppose that two agents with index χ_1 and χ_2 have the largest relative distance at time t, i.e., $\|x_{\chi_1}(t) - x_{\chi_2}(t)\|_2 = \max_{i,j \in \Gamma} \|x_i(t) - x_j(t)\|_2$. Then it can be obtained that $\|e_i(t)\|_2 \leq \frac{1}{N} \sum_{j=1}^{N} \|x_i(t) - x_j(t)\|_2 \leq \|x_{\chi_1}(t) - x_{\chi_2}(t)\|_2$. Since the graph is assumed to be connected, $\|x_{\chi_1}(t) - x_{\chi_2}(t)\|_2 \leq \sum_{i=1}^{N} \sum_{j=1}^{N} \|x_i(t) - x_j(t)\|_2$. Therefore, one has that

$$
\begin{aligned}
\|e_i(t)\|_2 &\leq \sum_{i=1}^{N} \sum_{j=1}^{N} \|x_i(t) - x_j(t)\|_2 \\
&= \sum_{i=1}^{N} \sum_{j=1}^{N} \|e_i(t) - e_j(t)\|_2
\end{aligned}
$$

$$\leq \sum_{i=1}^{N}\sum_{j=1}^{N} \|e_i(t) - e_j(t)\|_1. \tag{7.13}$$

Furthermore, one obtains

$$\begin{aligned}
-k_1 \sum_{i=1}^{N} e_i^T \varphi_i(x_i) \leq & \sum_{i=1}^{N} \left\|e_i\right\|_2 \left\|k_1 \varphi_i(x_i)\right\|_2 \\
\leq & N k_1 g_0 \sum_{i=1}^{N}\sum_{j=1}^{N} \left\|e_i - e_j\right\|_1
\end{aligned} \tag{7.14}$$

for the fourth term in (7.7). Combining (7.8)–(7.14), it can be concluded from (7.7) that

$$\begin{aligned}
\dot{V} \leq & -\frac{\alpha_1}{2}\sum_{i=1}^{N}\sum_{j=1}^{N} a_{ij}\left\|e_i - e_j\right\|_{2-\frac{a}{b}}^{2-\frac{a}{b}} \\
& -\frac{\beta_1}{2}\sum_{i=1}^{N}\sum_{j=1}^{N} a_{ij}\left\|e_i - e_j\right\|_{2+\frac{a}{b}}^{2+\frac{a}{b}} - k_1\rho\sum_{i=1}^{N}\left\|e_i\right\|_2^2 - \frac{\gamma_1}{2}\sum_{i=1}^{N}\sum_{j=1}^{N} a_{ij}\left\|e_i - e_j\right\|_1 \\
& + N k_1 g_0 \sum_{i=1}^{N}\sum_{j=1}^{N}\left\|e_i - e_j\right\|_1 \\
\leq & -\frac{\alpha_1}{2}\sum_{i=1}^{N}\sum_{j=1}^{N} a_{ij}\left\|e_i - e_j\right\|_{2-\frac{a}{b}}^{2-\frac{a}{b}} - \frac{\beta_1}{2}\sum_{i=1}^{N}\sum_{j=1}^{N} a_{ij}\left\|e_i - e_j\right\|_{2+\frac{a}{b}}^{2+\frac{a}{b}} \\
\leq & -\frac{\alpha_1}{2}\sum_{i=1}^{N}\sum_{j=1}^{N} a_{ij}\left\|e_i - e_j\right\|_2^{2-\frac{a}{b}} - \frac{\beta_1}{2}\sum_{i=1}^{N}\sum_{j=1}^{N} a_{ij} m^{-\frac{a}{2b}}\left\|e_i - e_j\right\|_2^{2+\frac{a}{b}} \\
= & -\frac{\alpha_1}{2}\sum_{i=1}^{N}\sum_{j=1}^{N}\left(a_{ij}^{\frac{2b}{2b-a}}\left\|e_i - e_j\right\|_2^2\right)^{1-\frac{a}{2b}} \\
& -\frac{\beta_1}{2} m^{-\frac{a}{2b}}\sum_{i=1}^{N}\sum_{j=1}^{N}\left(a_{ij}^{\frac{2b}{2b+a}}\left\|e_i - e_j\right\|_2^2\right)^{1+\frac{a}{2b}} \\
\leq & -\frac{\alpha_1}{2}\left(\sum_{i=1}^{N}\sum_{j=1}^{N} a_{ij}^{\frac{2b}{2b-a}}\left\|e_i - e_j\right\|_2^2\right)^{1-\frac{a}{2b}} \\
& -\frac{\beta_1}{2} m^{-\frac{a}{2b}} N^{-\frac{a}{b}}\left(\sum_{i=1}^{N}\sum_{j=1}^{N} a_{ij}^{\frac{2b}{2b+a}}\left\|e_i - e_j\right\|_2^2\right)^{1+\frac{a}{2b}}.
\end{aligned} \tag{7.15}$$

Let $G_B = \sum_{i,j=1}^{N} a_{ij}^{2b/(2b-a)} \|e_i - e_j\|_2^2$ and $G_C = \sum_{i,j=1}^{N} a_{ij}^{2b/(2b+a)} \|e_i - e_j\|_2^2$. Note that $G_B = 2e^T(\mathscr{L}_B \otimes I_m)e$ and $G_C = 2e^T(\mathscr{L}_C \otimes I_m)e$, where $e = [e_1, e_2, \ldots, e_N]^T$. Moreover, $G_B \geq 2\lambda_2(\mathscr{L}_B)e^T e$ and $G_C \geq 2\lambda_2(\mathscr{L}_C)e^T e$. Therefore,

$$
\begin{aligned}
\dot{V} &\leq -\frac{\alpha_1}{2}\left(\frac{G_B}{V}V\right)^{1-\frac{a}{2b}} - \frac{\beta_1}{2}m^{-\frac{a}{2b}}N^{-\frac{a}{b}}\left(\frac{G_C}{V}V\right)^{1+\frac{a}{2b}} \\
&\leq -\frac{\alpha_1}{2}(4\lambda_2(\mathscr{L}_B)V)^{1-\frac{a}{2b}} - \frac{\beta_1}{2}m^{-\frac{a}{2b}}N^{-\frac{a}{b}}(4\lambda_2(\mathscr{L}_C)V)^{1+\frac{a}{2b}} \\
&= -\frac{\alpha_1}{2}(4\lambda_2(\mathscr{L}_B))^{1-\frac{a}{2b}}V^{1-\frac{a}{2b}} - \frac{\beta_1}{2}m^{-\frac{a}{2b}}N^{-\frac{a}{b}}(4\lambda_2(\mathscr{L}_C))^{1+\frac{a}{2b}}V^{1+\frac{a}{2b}}.
\end{aligned}
$$

By using Theorem 2.4 and *Comparison Principle*, it can be concluded that $V = 0$ with a settling time T. In particular, $T \leq T_1$. Since $V = 0, e_i = 0, \forall i \in \Gamma$. Therefore, the consensus is achieved in a fixed time, which completes the proof. □

Theorem 7.1 *For a multi-agent system (7.1), the communication topology $\mathscr{G}$ is connected and fixed, and Assumption 7.2.1 holds. If $\gamma_1 \geq 2Nk_1 g_0/a_0$, then the protocol (7.5) enables the agent state to achieve consensus in a fixed time T_1, and minimizes the overall cost function (7.2) as $t \to \infty$.*

Proof From Lemma 7.2, for $t \geq T_1$, the states of all agents reach a consensus, i.e., $x_i(t) = \frac{1}{N}\sum_{j=1}^{N} x_j(t) \triangleq x_0(t), \forall i \in \Gamma$. Then one has that

$$
\begin{aligned}
\dot{x}_0(t) &= \frac{1}{N}\sum_{i=1}^{N}\Big\{-\alpha_1\sum_{j=1}^{N}a_{ij}(x_i - x_j)^{1-\frac{a}{b}} - \beta_1\sum_{j=1}^{N}a_{ij}(x_i - x_j)^{1+\frac{a}{b}} \\
&\quad - \gamma_1\sum_{j=1}^{N}a_{ij}\text{sign}(x_i - x_j) - k_1\nabla f_i(x_i)\Big\} \\
&= \frac{1}{N}\sum_{i=1}^{N}(-k_1\nabla f_i(x_i)) = -\frac{k_1}{N}\sum_{i=1}^{N}\nabla f_i(x_0),
\end{aligned}
$$

where the symmetric property of undirected graphs is used. Moreover, one obtains that

$$
\begin{aligned}
\frac{\mathrm{d}\sum_{i=1}^{N} f_i(x_0)}{\mathrm{d}t} &= \sum_{i=1}^{N}\nabla f_i^T(x_0)\frac{\mathrm{d}x_0}{\mathrm{d}t} \\
&= -\frac{k_1}{N}\sum_{i=1}^{N}\nabla f_i^T(x_0)\sum_{j=1}^{N}\nabla f_j(x_0) \\
&= -\frac{k_1}{N}\left(\sum_{i=1}^{N}\nabla f_i(x_0)\right)^T\sum_{j=1}^{N}\nabla f_j(x_0).
\end{aligned}
\tag{7.16}
$$

By Assumption 7.2.1, $\sum_{i=1}^{N} f_i(x)$ is lower bounded. Therefore, it can be concluded that $\lim_{t\to\infty} \sum_{i=1}^{N} \nabla f_i(x_0) = 0$. Invoking Lemma 7.1, $x = x_0$ minimizes the differentiable convex function $\sum_{i=1}^{N} f_i(x)$ as $t \to \infty$. □

Due to the limited communication range of agents, there may exist some communication link failures or additions among networked agents. Accordingly, the communication topology may switch over time. In what follows, we investigate the switching topology case.

From the expression of V, it can be found that V is independent from the topology, so it is ready to extend the fixed topology results to the switching case. First, a switching signal $\varrho(t) : \mathbb{R}^+ \to \Theta$ is used to determine the graph property at any particular time, i.e., $\mathscr{G}(t) = \mathscr{G}_{\varrho(t)}$, where Θ is a finite set consisting of index numbers associated to specific connected graph patterns. For example, $\Theta = \{1, 2, \ldots, \kappa\}$ implies that there are κ different graph patterns. The graph is fixed in between any two consecutive switches, and the graph pattern changes from one to another indexed in Θ over time. Notations of $a_0(t)$, $\mathscr{L}_B(t)$, and $\mathscr{L}_C(t)$ are to denote a_0, the Laplacian matrix of graph $\mathscr{G}([a_{ij}^{2b/(2b-a)}])$, and $\mathscr{G}([a_{ij}^{2b/(2b+a)}])$ at time t, respectively. In addition, we define $\lambda_2^*(\mathscr{L}_B) = \min\{\lambda_2(\mathscr{L}_B(t))\}$, $\lambda_2^*(\mathscr{L}_C) = \min\{\lambda_2(\mathscr{L}_C(t))\}$, and $a_0^* = \min\{a_0(t)\}$.

Corollary 7.1 *For a multi-agent system (7.1), the communication graph is switching among Θ, i.e., $\mathscr{G}(t) = \mathscr{G}_{\varrho(t)}$, and Assumption 7.2.1 holds. If $\gamma_1 \geq 2Nk_1 g_0/a_0^*$, then the protocol (7.5) enables the agent state to achieve consensus in a fixed time of $\check{T}_1 = \frac{2\pi b m^{\frac{a}{4b}} N^{\frac{a}{2b}}}{a\sqrt{\alpha_1\beta_1}(4\lambda_2^*(\mathscr{L}_B))^{\frac{1}{2}-\frac{a}{4b}}(4\lambda_2^*(\mathscr{L}_C))^{\frac{1}{2}+\frac{a}{4b}}}$, and minimizes the overall cost function (7.2) as $t \to \infty$.*

Proof At time t, $\varrho(t) \in \Theta$, the time derivative of V is calculated as

$$\begin{aligned}
\dot{V} \leq & -\frac{\alpha_1}{2}(4\lambda_2(\mathscr{L}_B(t)))^{1-\frac{a}{2b}} V^{1-\frac{a}{2b}} \\
& -\frac{\beta_1}{2} m^{-\frac{a}{2b}} N^{-\frac{a}{b}} (4\lambda_2(\mathscr{L}_C(t)))^{1+\frac{a}{2b}} V^{1+\frac{a}{2b}} \\
\leq & -\frac{\alpha_1}{2}(4\lambda_2^*(\mathscr{L}_B))^{1-\frac{a}{2b}} V^{1-\frac{a}{2b}} \\
& -\frac{\beta_1}{2} m^{-\frac{a}{2b}} N^{-\frac{a}{b}} (4\lambda_2^*(\mathscr{L}_C))^{1+\frac{a}{2b}} V^{1+\frac{a}{2b}}.
\end{aligned} \tag{7.17}$$

Obviously, $\forall \varrho(t) \in \Theta$, (7.17) holds. In fact, V serves as a common Lyapunov function [5] for any $\varrho(t) \in \Theta$. Therefore, the stability of (7.1) is guaranteed under the switching topology when using protocol (7.5). Following the same line in Lemma 7.2, it is concluded that the fixed-time consensus is achieved, and the estimated upper bound of settling time is $\check{T}_1$. For $t \geq \check{T}_1$, the states of all agents will reach the agreement. Then, along the same line in Theorem 7.1, the overall cost function is minimized. □

7.3 Distributed Optimization with Time-Varying Cost Functions

In this section, we investigate the time-varying distributed optimization problem (7.3) for multi-agent systems (7.1). Note that (7.3) is equivalent to the problem that all agents reach a consensus while minimizing the overall cost function $\sum_{i=1}^{N} f_i(x_i, t)$, i.e.,

$$\text{minimize} \sum_{i=1}^{N} f_i(x_i, t), \quad \text{subject to} \quad x_i = x_j \in \mathbb{R}^m. \tag{7.18}$$

Based on protocol (7.5), we propose an edge-based distributed protocol for the ith agent, i.e.,

$$\begin{aligned} u_i = &-\alpha_2 \sum_{j=1}^{N} a_{ij}(x_i - x_j)^{1-\frac{a}{b}} - \beta_2 \sum_{j=1}^{N} a_{ij}(x_i - x_j)^{1+\frac{a}{b}} \\ &- \gamma_2 \sum_{j=1}^{N} a_{ij}\text{sign}(x_i - x_j) + \eta_i, \end{aligned} \tag{7.19}$$

where $\eta_i = -H_i^{-1}(x_i, t)\big(\nabla f_i(x_i, t) + \frac{\partial \nabla f_i(x_i,t)}{\partial t}\big)$.

Assumption 7.3.1 For the cost function $f_i(x, t)$, it is twice continuously differentiable with respect to $x \in \mathbb{R}^m$, and the Hessian matrix $H_i(x, t) = H_j(x, t), \forall i, j \in \Gamma$. In addition, $H_i(x, t)$ is invertible for $\forall x, t$, and the term η_i can be further rewritten in a form of $\zeta x_i + \vartheta_i(t)$, where ζ is a nonpositive constant and $\|\vartheta_i(t)\|_2$ is upper bounded by a positive constant ϕ_0.

Remark 7.3 Compared to protocol (7.5), the difference of protocol (7.19) is the term of η_i, which takes the Hessian matrix of time-varying functions into account. In fact, if the time-varying function $f_i(x_i, t)$ is reduced to the time-invariant function $f_i(x_i)$, then $\frac{\partial \nabla f_i(x_i,t)}{\partial t} = 0$ in η_i. In this case, protocol (7.19) can be reduced to the form of (7.5).

Lemma 7.3 *For a multi-agent system (7.1), the communication topology $\mathcal{G}$ is connected and Assumption 7.3.1 holds. If $\gamma_2 \geq 2N\phi_0/a_0$, then the protocol (7.19) enables the system to achieve fixed-time consensus. Moreover, the settling time $T \leq T_2$, and*

$$T_2 = \frac{2\pi b m^{\frac{a}{4b}} N^{\frac{a}{2b}}}{a\sqrt{\alpha_2 \beta_2}(4\lambda_2(\mathcal{L}_B))^{\frac{1}{2}-\frac{a}{4b}}(4\lambda_2(\mathcal{L}_C))^{\frac{1}{2}+\frac{a}{4b}}}. \tag{7.20}$$

Proof The proof is similar to that of Lemma 7.2, so we only present the different part. Consider the same Lyapunov function candidate $V = \frac{1}{2}\sum_{i=1}^{N} e_i^T e_i$, and its time derivative is

$$\begin{aligned}
\dot{V} &= \sum_{i=1}^{N} e_i^T \dot{e}_i \\
&= \sum_{i=1}^{N} e_i^T \Bigg(-\alpha_2 \sum_{j=1}^{N} a_{ij}(x_i - x_j)^{1-\frac{a}{b}} - \frac{1}{N} \sum_{j=1}^{N} \dot{x}_j - \beta_2 \sum_{j=1}^{N} a_{ij}(x_i - x_j)^{1+\frac{a}{b}} + \eta_i \\
&\quad - \gamma_2 \sum_{j=1}^{N} a_{ij} \mathrm{sign}(x_i - x_j) \Bigg). \qquad (7.21)
\end{aligned}$$

Since η_i can be further rewritten in a form of $\zeta x_i + \vartheta_i(t)$, similar to (7.14), one gets

$$\sum_{i=1}^{N} e_i^T \vartheta_i(t) \leq \sum_{i=1}^{N} \left\| e_i \right\|_2 \left\| \vartheta_i(t) \right\|_2 \leq N\phi_0 \sum_{i=1}^{N} \sum_{j=1}^{N} \left\| e_i - e_j \right\|_1, \qquad (7.22)$$

and similar to (7.9), one has

$$\sum_{i=1}^{N} e_i^T \zeta x_i = \zeta \sum_{i=1}^{N} \left\| e_i \right\|_2^2. \qquad (7.23)$$

Combining (7.22), (7.23) with (7.10), (7.11), and (7.12), one finally obtains

$$\begin{aligned}
\dot{V} \leq &- \frac{\alpha_2}{2} (4\lambda_2(\mathscr{L}_B))^{1-\frac{a}{2b}} V^{1-\frac{a}{2b}} \\
&- \frac{\beta_2}{2} m^{-\frac{a}{2b}} N^{-\frac{a}{b}} (4\lambda_2(\mathscr{L}_C))^{1+\frac{a}{2b}} V^{1+\frac{a}{2b}}, \qquad (7.24)
\end{aligned}$$

where $\gamma_2 \geq 2N\phi_0/a_0$ is used. Then similarly to the proof of Lemma 7.2, the system achieves the fixed-time consensus within T_2. □

Remark 7.4 While the requirement under Assumption 7.3.1 that all agents have the same Hessian matrix and $\eta_i = \zeta x_i + \vartheta_i(t)$ can be restrictive, there exists an important class of cost functions that satisfy both requirements, i.e., the objective functions that are commonly used for energy minimization [6]. For example, $f_i(x_i, t) = (cx_i + g_i(t))^2$, where $c \in \mathbb{R}$, $g_i(t)$ is a time-varying function. In Section V, we will provide a detailed example that satisfies Assumption 7.3.1.

Theorem 7.2 *For a multi-agent system (7.1), the communication topology $\mathscr{G}$ is connected, and Assumption 7.3.1 holds. Then the protocol (7.19) with $\gamma_2 \geq 2N\phi_0/a_0$ guarantees that the state achieves consensus in a fixed time T_2, and minimizes the overall cost function (7.3) as $t \to \infty$.*

Proof From Lemma 7.3, for $t \geq T_2$, the states of all agents reach a consensus, and thus the optimization problem (7.3) becomes (7.18). When $t \geq T_2$, suppose that the consensus state is x_0. In this case, $\dot{x}_0 = \eta_i$. Construct a Lyapunov function candidate $V_0 = \frac{1}{2}\left(\sum_{i=1}^{N} \nabla f_i(x_0, t)\right)^T \left(\sum_{i=1}^{N} \nabla f_i(x_0, t)\right)$, and its time derivative is

$$
\begin{aligned}
\dot{V}_0 =& \Big(\sum_{i=1}^{N} \nabla f_i(x_0, t)\Big)^T \Big(\sum_{i=1}^{N} H_i(x_0, t)\dot{x}_0 + \sum_{i=1}^{N} \frac{\partial \nabla f_i(x_0, t)}{\partial t}\Big) \\
=& \Big(\sum_{i=1}^{N} \nabla f_i(x_0, t)\Big)^T \Big(H_i(x_0, t)\Big(\sum_{i=1}^{N} \dot{x}_0 + H_i^{-1}(x_0, t) \sum_{i=1}^{N} \frac{\partial \nabla f_i(x_0, t)}{\partial t}\Big)\Big) \\
=& -\Big(\sum_{i=1}^{N} \nabla f_i(x_0, t)\Big)^T \Big(\sum_{i=1}^{N} \nabla f_i(x_0, t)\Big),
\end{aligned} \tag{7.25}
$$

where the assumption that $H_i(x_i, t) = H_j(x_j, t)$, $\forall i, j \in \Gamma$, is used to obtain the second equality. Since V_0 is positive definite if $\nabla f_i(x_0, t) \neq 0$, it can be concluded that $\lim_{t \to \infty} \sum_{i=1}^{N} \nabla f_i(x_0, t) = 0$. Invoking Lemma 7.1, $x = x_0$ minimizes the differentiable convex function $\sum_{i=1}^{N} f_i(x, t)$ as $t \to \infty$, which completes the proof. □

For the switching communication topology, similar to the proof of Corollary 7.1, we have the following corollary.

Corollary 7.2 *For a multi-agent system (7.1), the communication topology $\mathcal{G}$ is switching among Θ, $\mathcal{G}(t) = \mathcal{G}_{\varrho(t)}$, and Assumption 7.3.1 holds. If $\gamma_2 \geq 2N\phi_0/a_0^*$, then the protocol (7.19) enables the agent state to achieve consensus in a fixed time $\check{T}_2 = \frac{2\pi b m^{\frac{a}{4b}} N^{\frac{a}{2b}}}{a\sqrt{\alpha_2\beta_2}(4\lambda_2^*(\mathcal{L}_B))^{\frac{1}{2}-\frac{a}{4b}}(4\lambda_2^*(\mathcal{L}_C))^{\frac{1}{2}+\frac{a}{4b}}}$, and minimizes the overall cost function (7.3) as $t \to \infty$.*

7.4 Numerical Examples

In this section, we investigate a package-level state-of-charge (SOC) balancing problem for a grid-connected battery energy storage system (BESS). The BESS plays an important role in energy storage systems, such as providing energy buffers to cooperate with renewable energy sources [7]. A BESS may consist of multiple battery packages and each package consists of a number of battery cells. For each package, a battery management system (BMS) is used to balance the SOCs of all the cells to maintain high efficiency [8]. Additionally, a package-level SOC balancing can provide safety redundancy in protecting batteries from overcharged or discharged [9].

The relationship, for the ith battery package, between the power output P_i and the SOC can be expressed as [10]

$$
\frac{\mathrm{d}SOC_i}{\mathrm{d}t} \approx -\frac{EFF_i}{QV_d} P_i, \tag{7.26}
$$

where EFF_i is determined by the coulombic efficiency [11] and the converter conversion efficiency [12] of the ith battery package, Q denotes the capacity of each battery package, and V_d denotes a battery related constant. The aim is to balance the SOC_i for all battery packages. A distributed control scheme proposed in [10] is modified to achieve the goal. The block diagram is shown in Fig. 7.2, where v_i and i_i denotes the output voltage and current of the ith converter, respectively, v denotes the voltage of the distribution bus, c_i is an auxiliary variable for the energy coordinator, P^* is the final converged power output value, and $\hat{p}_i$ denotes the adjusted local reference power output for the ith battery package.

In what follows, we focus on achieving the power output consensus while optimizing the local cost, i.e., obtaining P^* for each package, by using the distributed optimization protocols proposed in this chapter. Three examples are provided to illustrate the effectiveness of the theoretical analysis. Without loss of generality, it is assumed that there are six battery packages, i.e., $N = 6$, for the three examples, and P_i can be decomposed into a two-dimensional Cartesian coordinate system, i.e., $m = 2$. In addition, the communication topology used in Examples 1 and 2 is fixed, shown in Fig. 7.1.

Example 1: Distributed optimization with time-invariant cost functions. The time-invariant local cost function for the ith battery package is chosen as $f_i(P_i) = (P_{i1} + 0.1i)^2 + (P_{i2} - 0.1i)^2$. It can be easily calculated that the minima of the overall cost function $\sum_{i=1}^{6} f_i(P_i)$ is 0.35. Set the initial states of six battery packages as (2, 1), (−3, 3), (1, 2), (−2, −3), (−1, 2), (2, 1). Select $\alpha_1 = \beta_1 = 3, k_1 = 2, a = 2, b = 5$. Under Assumption 7.2.1 and the expression of $f_i(x_i)$, it can be obtained that $g_0 = 1.7$. Therefore, $\gamma_1 = 2nk_1g_0/a_0 = 40.8$. Furthermore, $T_1 = 2.01$ s by Lemma 7.2. Implementing the protocol (7.5) with $N = 6$, the simulation results are shown in Figs. 7.3 and 7.4. Both figures show that the settling time is about $t = 0.05$ s, which is smaller than T_1. The evolution of $\sum_{i=1}^{6} f_i(P_i)$ is shown in Fig. 7.5, which finally approaches to the optimal value 0.35.

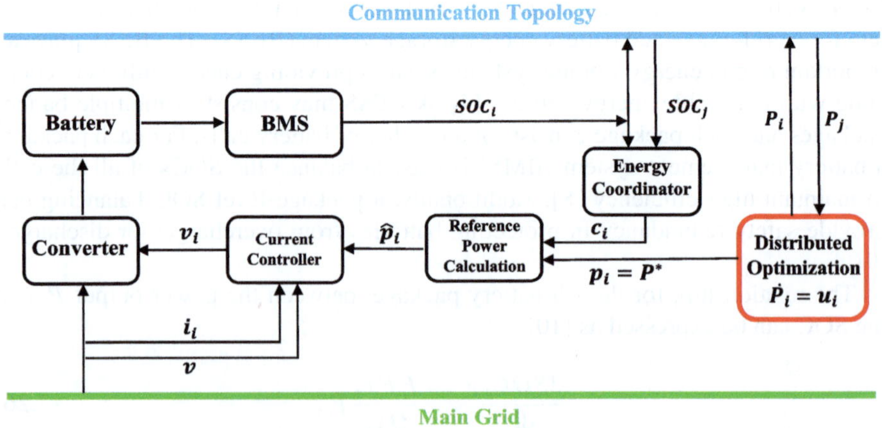

Fig. 7.1 A fixed communication topology

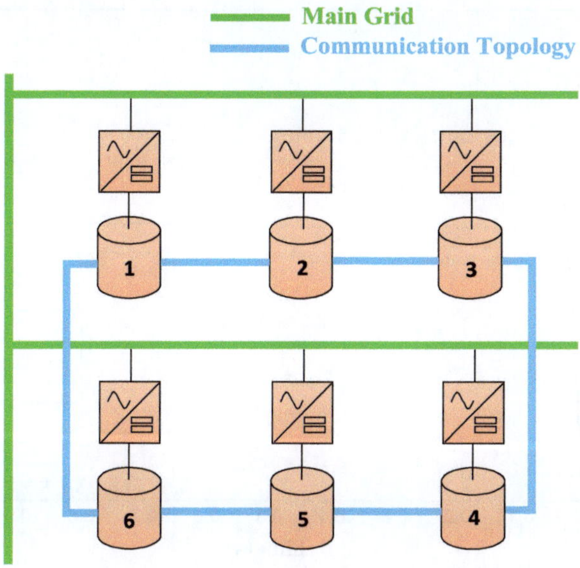

Fig. 7.2 The distributed control scheme for battery energy storage system

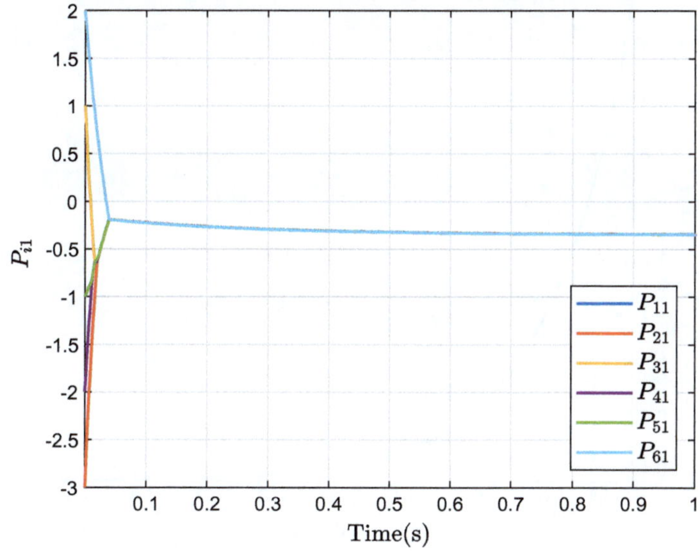

Fig. 7.3 State P_{i1} for time-invariant cost functions under protocol (7.5) with $N = 6$

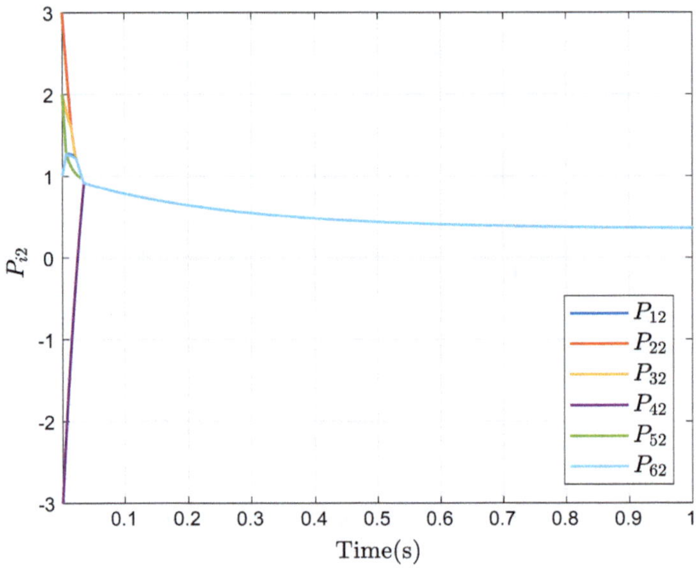

Fig. 7.4 State P_{i2} for time-invariant cost functions under protocol (7.5) with $N = 6$

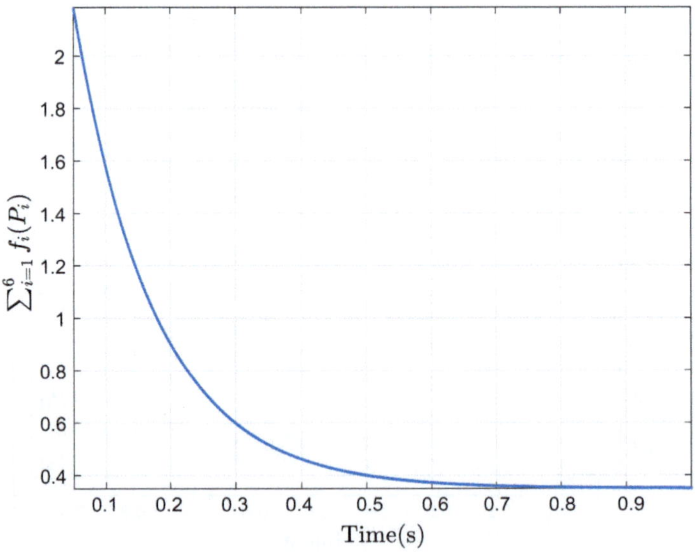

Fig. 7.5 The overall time-invariant cost functions under protocol (7.5) with $N = 6$

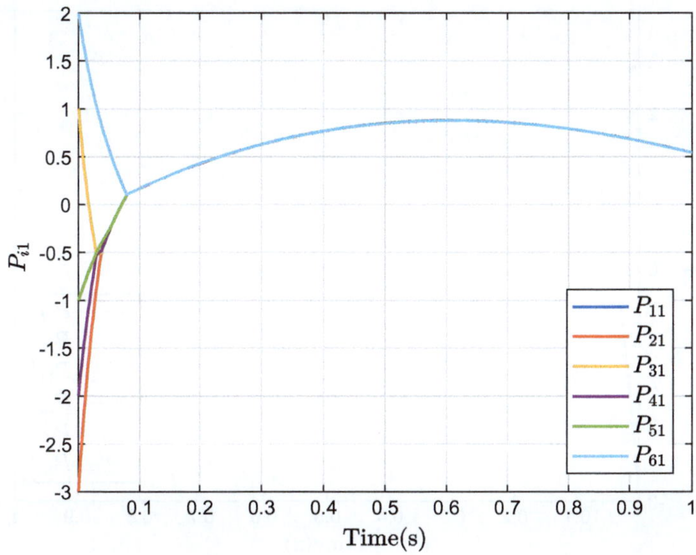

Fig. 7.6 State P_{i1} for time-varying cost functions under protocol (7.19) with $N = 6$

Example 2: Distributed optimization with time-varying cost functions. The time-varying local cost function for the ith battery package is chosen as $f_i(P_i, t) = (P_{i1} - i\cos(t))^2 + (P_{i2} - i\sin(t))^2$. It can be calculated that $\nabla f_i(P_i, t) = [2P_{i1} - 2i\cos(t);\ 2P_{i2} - 2i\sin(t)]$, $H_i^{-1}(P_i, t) = [0.5\ \ 0;\ 0\ \ 0.5]$, $\frac{\partial \nabla f_i(P_i,t)}{\partial t} = [2i\sin(t); -2i\cos(t)]$. Thus,

$$\eta_i = -\begin{bmatrix} P_{i1} \\ P_{i2} \end{bmatrix} + i \begin{bmatrix} \cos(t) - \sin(t) \\ \cos(t) + \sin(t) \end{bmatrix}, \tag{7.27}$$

which satisfies Assumption 7.3.1 with $\zeta = -1$ and $\vartheta_i(t) = i \begin{bmatrix} \cos(t) - \sin(t) \\ \cos(t) + \sin(t) \end{bmatrix}$. In particular, $\phi_0 = 6\sqrt{2}$, thereby selecting $\gamma_2 = 2n\phi_0/a_0 = 16.97$. All other constants keep the same as those in Example 1, so $T_2 = T_1 = 2.01$ s. Implementing the protocol (7.19) with $N = 6$, in Figs. 7.6 and 7.7, the settling time is about $t = 0.1$ s, which is smaller than T_2. In Fig. 7.8, it shows trajectories of P_i for the six battery packages that finally converge to a common one, which is in fact a circle centered at the origin with a radius of 3.5. Such a trajectory is created by the time-varying optimal point of the overall cost function $\sum_{i=1}^{6} f_i(P_i, t)$.

Example 3: Distributed optimization with time-varying cost functions over switching communication topologies. This example is a switching topology version of Example 2, i.e., instead of using a fixed graph, the interaction graph is arbitrarily switching among $\mathscr{G}_1$, $\mathscr{G}_2$, $\mathscr{G}_3$ and $\mathscr{G}_4$ as shown in Fig. 7.9. It can be easily calculated that $\lambda_2^*(\mathscr{L}_B) = \lambda_2^*(\mathscr{L}_C) = \min\{1, 0.2679, 0.5858, 0.4859\} = 0.2679$. Other parameters keep the same as those in Example 2. According to Corollary 7.2, $\check{T}_2 = 7.4937$s.

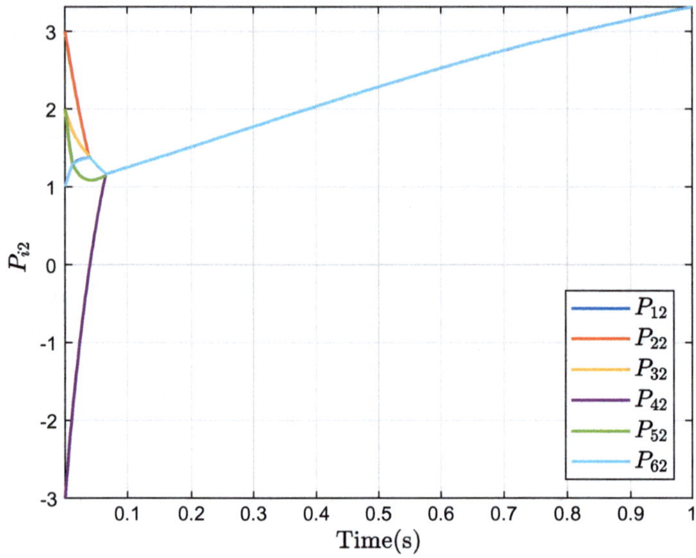

Fig. 7.7 State P_{i2} for time-varying cost functions under protocol (7.19) with $N = 6$

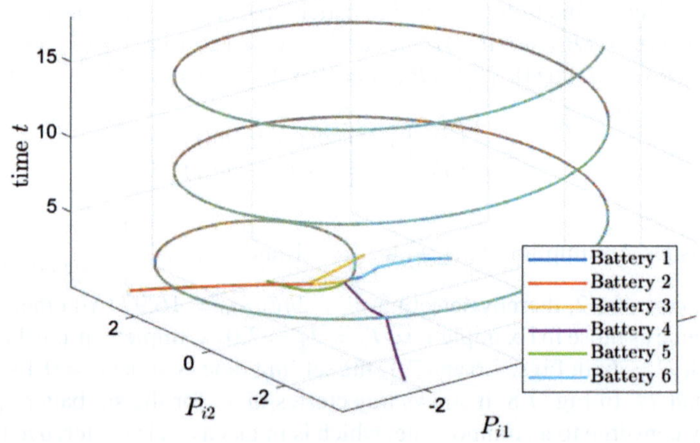

Fig. 7.8 Trajectories for time-varying cost functions under protocol (7.19) with $N = 6$

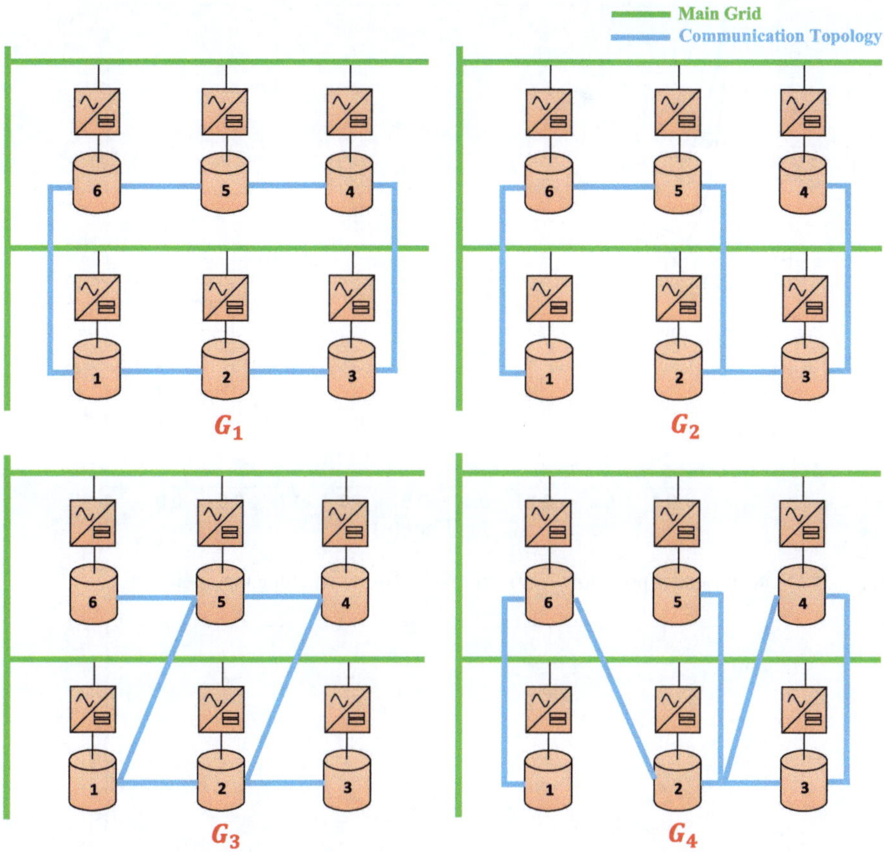

Fig. 7.9 Connected communication graphs

Setting the initial states as (12, 7), (−13, −9), (−8, 12), (9, −13), (−8, 8), (9, 7), and implementing (7.19) with $N = 6$, the result is shown in Figs. 7.10 and 7.11. The settling time is about $t = 0.2$ s, which demonstrates the correctness of $\check{T}_2$. In Fig. 7.12, trajectories of P_i for the six battery packages still converge to a dynamic circle centered at the origin with a radius of 3.5.

7.5 Conclusions

A distributed optimization problem for multi-agent systems with continuous single-integrator dynamics is investigated by using an edge-based fixed-time consensus strategy. For multi-agent systems with a class of time-invariant cost functions, a distributed protocol is first proposed to achieve the fixed-time consensus while minimizing the sum of convex cost functions known to the individual agents under both

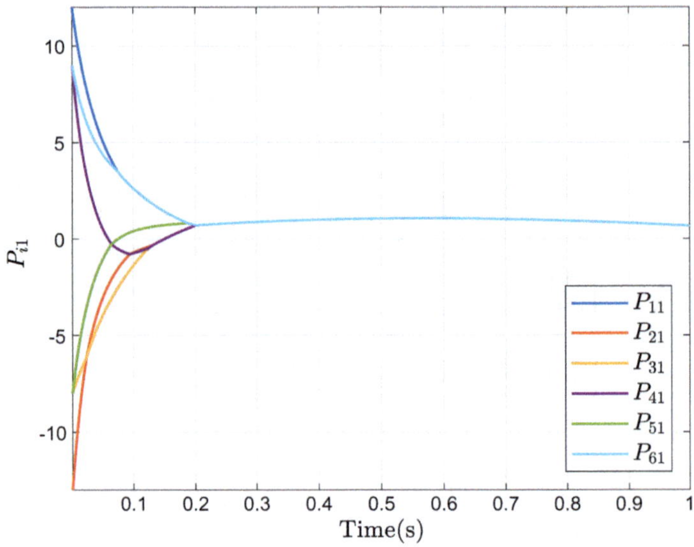

Fig. 7.10 State P_{i1} under protocol (7.19) with $N = 6$ over switching topologies

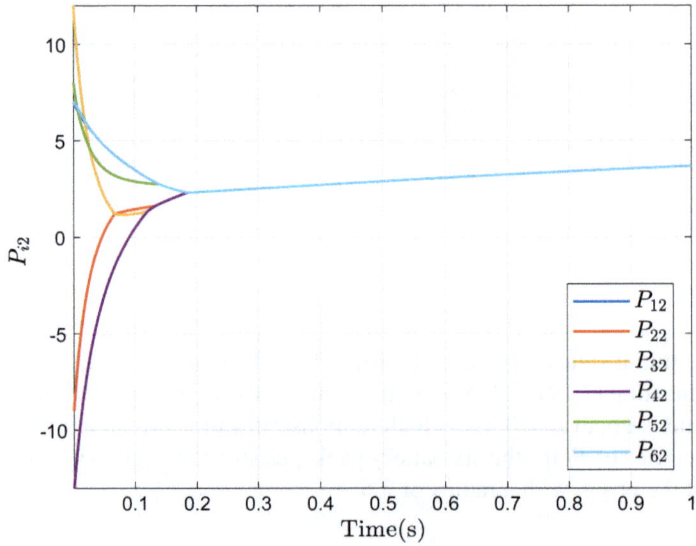

Fig. 7.11 State P_{i2} under protocol (7.19) with $N = 6$ over switching topologies

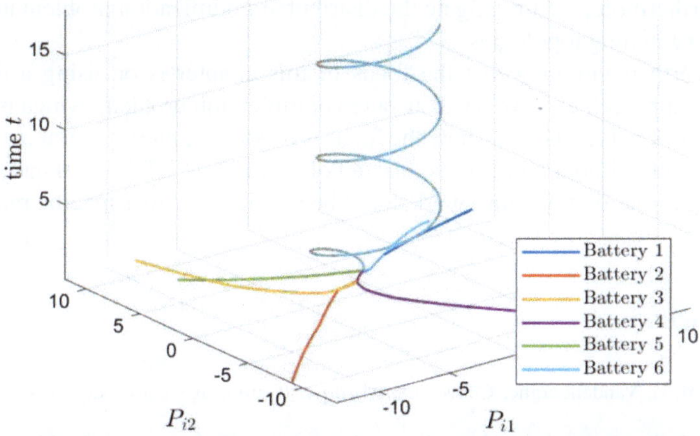

Fig. 7.12 Trajectories under protocol (7.19) with $N = 6$ over switching topologies

fixed and switching communication topologies. Then the results are extended to the case with a class of time-varying cost functions. Different from conventional strategies for distributed optimization, the current work obtains an upper bound of the settling time for consensus without dependence on initial conditions. The effectiveness of the proposed protocols is illustrated in case studies, which demonstrates the superiority of the proposed approach in a power output agreement for battery packages.

7.6 Notes

Different from the scheme in [10], a distributed optimization block (the red block in Fig. 7.2) is added in this chapter. Instead of assuming that P^* is given by a global command generator, we consider that P^* is obtained through a distributed optimization process. In other words, each package can actively determine its desired power output based on its local optimal strategy, which provides additional flexibility for the overall BESS. Once P^* is obtained by each package, the distributed control scheme in [10] can be directly applied to achieve the package-level SOC balancing. The results in this chapter are based on [13]. In [6], a time-varying distributed convex optimization problem is considered for systems with single/double-integrator dynamics by nonsmooth algorithms. The results in [6] are obtained under a fixed communication topology, and the state agreement is exponentially achieved over an infinite time horizon, or achieved in a finite time by using a homogeneity technique without providing an explicit bound for the settling time. In contrast, we achieve the fixed-time state agreement by using an edge-based fixed-time consensus approach, and derive an upper bound for the state agreement that is not dependent on initial states. Furthermore, we investigate the distributed optimization problem under both fixed and switching topologies.

It is worth mentioning that the focus of this chapter is on using a fixed-time consensus approach to solve the distributed optimization problem, which is different from the work in [4, 14–19], where the focus is on solving more general optimization problems with/without constraints. Instead of letting $x \in \mathbb{R}^m$ in (7.2) and (7.3), one interesting direction for future work could be restricting x in a specific range.

References

1. S. Boyd, L. Vandenberghe, *Convex Optimization* (Cambridge University Press, Cambridge, 2004)
2. Z. Zuo, L. Tie, A new class of finite-time nonlinear consensus protocols for multi-agent systems. Int. J. Control **87**(2), 363–370 (2014)
3. Z. Zuo, L. Tie, Distributed robust finite-time nonlinear consensus protocols for multi-agent systems. Int. J. Syst. Sci. **47**(6), 1366–1375 (2016)
4. A. Nedić, A. Ozdaglar, P.A. Parrilo, Constrained consensus and optimization in multi-agent networks. IEEE Trans. Autom. Control **55**(4), 922–938 (2010)
5. D. Liberzon, A.S. Morse, Basic problems in stability and design of switched systems. IEEE Control Syst. Mag. **19**(5), 59–70 (1999)
6. S. Rahili, W. Ren, Distributed continuous-time convex optimization with time-varying cost functions. IEEE Trans. Autom. Control **62**(4), 1590–1605 (2017)
7. X. Li, D. Hui, X. Lai, Battery energy storage station (BESS)-based smoothing control of photovoltaic (PV) and wind power generation fluctuations. IEEE Trans. Sustain. Energy **4**(2), 464–473 (2013)
8. W. Huang, J.A.A. Qahouq, Energy sharing control scheme for state-of-charge balancing of distributed battery energy storage system. IEEE Trans. Ind. Electron. **62**(5), 2764–2776 (2015)
9. L. Maharjan, S. Inoue, H. Akagi, J. Asakura, State-of-charge (SOC)-balancing control of a battery energy storage system based on a cascade pwm converter. IEEE Trans. Power Electron. **24**(6), 1628–1636 (2009)
10. H. Cai, G. Hu, Distributed control scheme for package-level state-of-charge balancing of grid-connected battery energy storage system. IEEE Trans. Ind. Inf. **12**(5), 1919–1929 (2016)
11. R. Kaushik, I. Mawston, Coulombic efficiency of lead/acid batteries, particularly in remote-area power-supply (RAPS) systems. J. Power Sour. **35**(4), 377–383 (1991)
12. T. LaBella, W. Yu, J.-S. Lai, M. Senesky, D. Anderson, A bidirectional-switch-based wide-input range high-efficiency isolated resonant converter for photovoltaic applications. IEEE Trans. Power Electron. **29**(7), 3473–3484 (2014)
13. B. Ning, Q.-L. Han, Z. Zuo, Distributed optimization for multiagent systems: An edge-based fixed-time consensus approach. IEEE Trans. Cybern. **49**(1), 122–132 (2019)
14. D. Yuan, S. Xu, H. Zhao, Distributed primal-dual subgradient method for multiagent optimization via consensus algorithms. IEEE Trans. Syst. Man Cybern. Part B **41**(6), 1715–1724 (2011)
15. G. Shi, K.H. Johansson, Y. Hong, Reaching an optimal consensus: Dynamical systems that compute intersections of convex sets. IEEE Trans. Autom. Control **58**(3), 610–622 (2013)
16. Q. Liu, J. Wang, A second-order multi-agent network for bound-constrained distributed optimization. IEEE Trans. Autom. Control **60**(12), 3310–3315 (2015)
17. S.S. Kia, J. Cortés, S. Martínez, Distributed convex optimization via continuous-time coordination algorithms with discrete-time communication. Automatica **55**, 254–264 (2015)

18. Z. Qiu, S. Liu, L. Xie, Distributed constrained optimal consensus of multi-agent systems. Automatica **68**, 209–215 (2016)
19. S. Yang, Q. Liu, J. Wang, A multi-agent system with a proportional-integral protocol for distributed constrained optimization. IEEE Trans. Autom. Control **62**(7), 3461–3467 (2017)

Chapter 8
Distributed Optimization with Preserved Network Connectivity

This chapter is concerned with the distributed optimization of a multi-agent system with network connectivity preservation. In order to realize cooperative interactions, a connected network is the prerequisite for high-quality information exchange among agents. However, sensing or communication capability is range limited, so it is impractical to simply make an assumption that network connectivity is preserved by default. To address this concern, a class of generalized potentials including discontinuities caused by unexpected obstacles or noises are designed. For a class of quadratic cost functions, based on the potentials, a distributed protocol is proposed to formally guarantee the network connectivity over time and to realize the state agreement in a finite time while the sum of local functions known to individual agents is optimized. Since the right-hand side of the proposed protocol is discontinuous, some non-smooth analysis tools are applied to analyze the system performance. In some practical scenarios, where initial states are unavailable, a distributed protocol is further developed to realize the consensus in a fixed time while solving the distributed optimization problem and maintaining the network connectivity.

8.1 Problem Formulation

Consider a multi-agent system that consists of N agents labeled with index $1, 2, \ldots,$ N. Each agent evolves with the dynamics of

$$\dot{x}_i(t) = u_i(t), \quad i \in \Gamma, \tag{8.1}$$

where $x_i(t) \in \mathbb{R}^m$ and $u_i(t) \in \mathbb{R}^m$ denote the state and the control protocol, respectively, for the ith agent.

For the ith agent, it has a local cost function, denoted as $f_i(x) : \mathbb{R}^m \to \mathbb{R}$, which is only known to agent i itself. A conventional distributed optimization problem is to

Z. Zuo et al., *Fixed-Time Cooperative Control of Multi-Agent Systems*,
https://doi.org/10.1007/978-3-030-20279-8_8

design $u_i(t)$ by using local information together with neighboring interactions such that all agents finally arrive at the optimal state x^* solving a convex optimization problem of

$$\text{minimize} \quad \sum_{i=1}^{N} f_i(x), \quad \text{subject to} \quad x \in \mathbb{R}^m. \tag{8.2}$$

Note that (8.2) is equivalent to the problem that all agents reach a consensus while minimizing the overall cost function $\sum_{i=1}^{N} f_i(x_i)$, i.e.,

$$\text{minimize} \quad \sum_{i=1}^{N} f_i(x_i), \quad \text{subject to} \quad x_i = x_j \in \mathbb{R}^m. \tag{8.3}$$

Definition 8.1 For the system (8.1), if there exists a suitable protocol $u_i(t)$ such that

$$\begin{cases} \lim_{t \to T_f} \|x_i(t) - x_j(t)\| = 0; \\ \|x_i(t) - x_j(t)\| = 0, \quad \forall t \geq T \end{cases}$$

for all $i, j \in \Gamma$, where T is dependent on initial states, finite-time consensus is achieved. Furthermore, if T is upper bounded by $T_{\max}$ without dependence on initial states, fixed-time consensus is achieved.

In this chapter, a range R is used to mimic the limited communication capability of agents, i.e., no direct information exchange exists for any two agents with a distance larger than R. Under this limited range R, the problem to be addressed is, for the system (8.1), to design a suitable $u_i(t)$ such that

- network connectivity is preserved over time, i.e., there always exists a communication flow[1] for any two agents;
- finite-time consensus or fixed-time consensus is realized; and
- the cost function $\sum_{i=1}^{N} f_i(x)$ in (8.2) is minimized.

Assumption 8.1.1 For the cost function $f_i(x)$, it is quadratic with a positive-definite Hessian matrix. In particular, $\nabla f_i(x) = \rho x + \varphi$ holds for $\forall x \in \mathbb{R}^m$, where $\rho > 0$ and $\varphi \in \mathbb{R}^m$. □

Remark 8.1 Under Assumption 8.1.1, the requirement of $\nabla f_i(x) = \rho x + \varphi$ indicates that $f_i(x)$ belongs to a class of quadratic functions. As a first attempt to investigate distributed optimization taking both connectivity preservation and finite-time consensus into account, we concentrate on the quadratic-type functions. For more general forms of $f_i(x)$, a detailed discussion is provided in Remark 8.8.

---

[1]The communication flow consists of a single link or multiple consecutive links. Each link has a length smaller than R. Moreover, we consider undirected links, i.e., the interaction between two connected agents is symmetrical. If the link becomes directed, the detail-balanced condition is required to obtain the results developed in this chapter.

Some non-smooth analysis tools, definitions, and preliminary lemmas are provided in this subsection.

Definition 8.2 ([1, 2]) Consider the following differential equation, where the right-hand side can be discontinuous, i.e.,

$$\dot{x} = F(x), \tag{8.4}$$

where $F : \mathbb{R}^m \to \mathbb{R}^m$ is measurable and essentially locally bounded, and then $x(\cdot)$ is called a Filippov solution of (8.4) on $[t_0, t_1] \subset \mathbb{R}$ if it is absolutely continuous and it satisfies the differential inclusion $\dot{x} \in \mathscr{K}[F](x)$, for almost all $t \in [t_0, t_1]$, where $\mathscr{K}[F](x)$ is a set-valued map defined as

$$\mathscr{K}[F](x) = \overline{\text{co}}\{\lim F(\hat{x}) \mid \hat{x} \to x, \hat{x} \notin M\},$$

where $\overline{\text{co}}$ and M denote the convex closure and the set of measure zero, respectively.

Definition 8.3 ([3]) Let $\mathrm{f} : \mathbb{R}^m \to \mathbb{R}$ be a locally Lipschitz function. Then the set-valued Lie derivative of f with respect to F at x is

$$\tilde{\mathscr{L}}_F \mathrm{f}(x) = \{\varrho \in \mathbb{R} \mid \exists \varpi \in \mathscr{K}[F](x) \text{ such that } \eta^T \varpi = \varrho, \forall \eta \in \partial \mathrm{f}(x)\},$$

where $\partial \mathrm{f}(x)$ denotes the generalized gradient of f, which is defined by

$$\partial \mathrm{f}(x) = \text{co}\{\lim_{k\to\infty} \nabla \mathrm{f}(x_k) \mid x_k \to x, x_k \notin M \cup \Omega_\mathrm{f}\},$$

where co is the convex hull, and Ω_f denotes the set of points where f is not differentiable.

8.2 Distributed Optimization with Network Connectivity and Finite-Time Consensus

In this section, we design a class of generalized potentials with discontinuities. Based on the potentials, we propose a distributed protocol such that problem (8.2) is solved with both network connectivity maintenance and finite-time consensus.

8.2.1 Selection of the Communication Range

In the communication society, a metric of a bit error rate (BER) is commonly used to indicate the communication quality from one agent to another [4]. The BER, namely P_b, is defined as

$$P_b = \text{Prob}\{\pi \neq \hat{\pi}\}, \tag{8.5}$$

where π and $\hat{\pi}$ denote a transmitted bit and a received bit, respectively. Under M-QAM modulation (commonly used modulation in the communication literature), the BER is approximately characterized as [4]

$$P_b \approx 0.2e^{-1.5\frac{\sigma}{M-1}}, \tag{8.6}$$

where M denotes a modulation-related parameter, and σ denotes the received signal-to-noise ratio (SNR), i.e.,

$$\sigma = \frac{\varepsilon}{\vartheta^{\theta}}, \tag{8.7}$$

where ε is a parameter determined by the transmit power, antenna gain, and operation frequency, and ϑ is the relative distance between two communicating agents, θ is a path loss exponent (usually in the range of 2–6) depending on the environment.

The BER in (8.6) shows how the communication performance is affected by the received SNR, modulation, and other transmission parameters. Next, the probability of correct reception, namely P_c, from one agent to another is approximately characterized as

$$P_c \approx (1 - 0.2e^{-1.5\frac{\sigma}{M-1}}). \tag{8.8}$$

From the expression in (8.8), one can estimate a maximum distance $d_{\max}$ once a required minimum $P_{c_{\min}}$ is provided. Then the communication range R is selected as

$$R \leq d_{\max}. \tag{8.9}$$

Note that the higher $P_{c_{\min}}$ indicates that the selected communication range R becomes smaller, so the connectivity work needs to be carried out under this smaller R, which leaves less flexibility for cooperative interactions since sometimes agents need to keep a relatively large distance to avoid collisions. In other words, there exists a trade-off between the requirement for the communication quality and the preserved link length.

8.2.2 Generalized Potentials and Network Connectivity

We consider an artificial potential $\psi(s) : [0, \infty) \to [0, \infty)$ satisfying the following generalized properties:

- $\psi(s)$ is continuous on $[0, R)$, except on a set of isolated and countable points $\{\rho_k^i\}$ within $(0, R)$, i.e., there exist finite right and left limits, $\psi^+(\rho_k^i)$ and $\psi^-(\rho_k^i)$, respectively, $k = 1, 2, \ldots$ Moreover, $\psi(s)$ is discontinuous at point R;
- there exists a positive constant $\kappa \geq 1$ such that $\psi(s) \geq \kappa$ for $0 \leq s < R$ and $\psi(s) = 0$ for all $s \geq R$; and

- $\psi(s) \to +\infty$ when $s \to R$.

Instead of using a fixed a_{ij}, in this chapter, we use a state-dependent $a_{ij} = \psi(\|x_i - x_j\|)$ to model the connectivity strength between agents. More specifically, the following interaction rule is implemented to determine the connectivity strength:

$$a_{ij}(t=0) = \begin{cases} \psi(\|x_i(0) - x_j(0)\|), & \|x_i(0) - x_j(0)\| < R; \\ 0, & \|x_i(0) - x_j(0)\| \geq R, \end{cases}$$

and

$$a_{ij}(t>0) = \begin{cases} \psi(\|x_i(t) - x_j(t)\|), & (\|x_i(t) - x_j(t)\| < R) \cap \\ & (\|x_i(0) - x_j(0)\| < R); \\ 0, & \text{otherwise}. \end{cases}$$

How to select R is presented in Sect. 8.2.1. In this chapter, we aim to guarantee that there always exists a communication flow for any two agents. Note that a connected link between agents is the prerequisite for high-quality communication exchange.

Assumption 8.2.1 The initial network is connected, i.e., for any two agents at $t = 0$, there exists a path consisting of links, each of which has a positive connectivity strength $\psi(\cdot)$.

Remark 8.2 The requirement of the initial network connectivity under Assumption 8.2.1 is mild and common [5–8]. In fact, agents can be initially placed in a specific area to guarantee that the length of each link is smaller than R. Besides, the above interaction rule indicates that if the initial relative distance of two agents is not smaller than R, they will not interact with each other all the time. This is understood in the sense that two agents are "neighbors" does not necessary indicate they are "friends" [5]. In Sect. 8.2.4, we will prove that all initial connected links are preserved over time. As a result, the interaction graph is in fact fixed in this chapter.

Remark 8.3 In terms of $\psi(s)$, the first property is to consider unexpected obstacles or noises, which can cause discontinuities for the potential; the second property is to ensure that the connectivity strength is not too small, otherwise the convergence speed becomes slow; and the third property is to guarantee that the initial connected links are maintained over time. A formal proof about connectivity preservation is provided in Sect. 8.2.4.

8.2.3 Distributed Protocol

To solve the problem (8.2), we propose a distributed protocol for the ith agent, i.e.,

$$u_i = -\alpha \sum_{j=1}^{N} \psi(\|x_i - x_j\|) \lceil x_i - x_j \rfloor^{\mu} - \gamma \nabla f_i(x_i) \tag{8.10}$$

where $\alpha > 0$, $\gamma > 0$, and $0 < \mu < 1$.

Remark 8.4 In protocol (8.10), the first term aims to achieve the finite-time consensus, while the second term of $-\gamma \nabla f_i(x_i)$ is for the optimization purpose. Note that a state-dependent potential of $\psi(\|x_i - x_j\|)$ is integrated in the first term of (8.10), which is thus different from some conventional finite-time protocols [9–11]. According to the properties described in Sect. 8.2.2, $\psi(\|x_i - x_j\|)$ has discontinuities. Therefore, the right-hand side of (8.10) is discontinuous. In the sequel, some non-smooth techniques are exploited to analyze the system performance.

8.2.4 Performance Analysis

In this subsection, we first prove that initial connected links are preserved over time. Then we prove that all agents from a multi-agent system realize the state agreement in a finite time by using protocol (8.10) while the cost function $\sum_{i=1}^{N} f_i(x)$ in (8.2) is minimized. Construct the following nonnegative function:

$$V_1 = \frac{1}{2} \sum_{i,j=1}^{N} \int_0^{\|x_i - x_j\|} \alpha \psi(s) s^{\mu} \mathrm{d}s. \tag{8.11}$$

Theorem 8.1 *Under Assumption 8.1.1, for the system (8.1) with protocol (8.10), the network connectivity is preserved over time.*

Proof From Definition 8.2, combining (8.1) and (8.10), one obtains that, $i \in \Gamma$,

$$\dot{x}_i = -\alpha \sum_{j=1}^{N} \psi(\|x_i - x_j\|) \lceil x_i - x_j \rfloor^{\mu} - \gamma \nabla f_i(x_i) \triangleq F_i, \tag{8.12}$$

so $\dot{x}_i \in \mathscr{K}[F_i]$. Let $\partial_i V_1$ denote the ith component of ∂V_1, which is computed as

$$\partial_i V_1 = \alpha \sum_{j=1}^{N} \Phi(\|x_i - x_j\|) \lceil x_i - x_j \rfloor^{\mu},$$

where $\Phi(\|x_i - x_j\|) = \mathrm{co}\{\lim \psi(\|\bar{x}_i - \bar{x}_j\|) \mid \bar{x}_i \to x_i, \bar{x}_j \to x_j\}$.

Under the second property in Sect. 8.2.2, we consider two cases: (i) $\psi(\|x_i - x_j\|)$ is continuous at $\|x_i - x_j\|$; and (ii) $\psi(\|x_i - x_j\|)$ is discontinuous at $\|x_i - x_j\|$.

Case (i): In this case, $\mathrm{co}\{\lim \psi(\|\bar{x}_i - \bar{x}_j\|) \mid \bar{x}_i \to x_i, \bar{x}_j \to x_j\} = \psi(\|x_i - x_j\|)$, so $\Phi(\|x_i - x_j\|)$ is a nonnegative scalar.

Case (ii): In this case, $\mathrm{co}\{\lim \psi(\|\bar{x}_i - \bar{x}_j\|) \mid \bar{x}_i \to x_i, \bar{x}_j \to x_j\} = [\psi^-(\|x_i - x_j\|), \psi^+(\|x_i - x_j\|)]$, so $\Phi(\|x_i - x_j\|)$ is a nonnegative interval.

Therefore, one concludes that $\Phi(\|x_i - x_j\|)$ always belongs to a nonnegative interval.

For any $z \in \widetilde{\mathscr{L}}_{F_i} V_1$, following Definition 8.3, there exists a $w_i \in \mathscr{K}[F_i]$ such that $z = \sum_{i=1}^{N} \xi_i^T w_i$ for all $\xi_i \in \partial_i V_1$, $i \in \Gamma$. Note that for any $w_i \in \mathscr{K}[F_i]$, there always exists a $\psi_{ij} = \psi_{ji} \in \Phi(\|x_i - x_j\|)$ such that

$$w_i = -\alpha \sum_{j=1}^{N} \psi_{ij} \lceil x_i - x_j \rfloor^{\mu} - \gamma \rho x_i - \gamma \varphi. \tag{8.13}$$

Since $z = \sum_{i=1}^{N} \xi_i^T w_i$ for all $\xi_i \in \partial_i V_1$, $i \in \Gamma$, one can choose

$$\xi_i = \alpha \sum_{j=1}^{N} \psi_{ij} \lceil x_i - x_j \rfloor^{\mu} \in \partial_i V_1.$$

Therefore,

$$\begin{aligned} z &= -\sum_{i=1}^{N} \| \sum_{j=1}^{N} \alpha \psi_{ij} \lceil x_i - x_j \rfloor^{\mu} \|_2^2 + \sum_{i,j=1}^{N} \alpha \psi_{ij} \left(\lceil x_i - x_j \rfloor^{\mu} \right)^T (-\gamma \rho x_i - \gamma \varphi) \\ &= -\sum_{i=1}^{N} \| \sum_{j=1}^{N} \alpha \psi_{ij} \lceil x_i - x_j \rfloor^{\mu} \|_2^2 - \frac{1}{2} \sum_{i,j=1}^{N} \alpha \gamma \rho \psi_{ij} \|x_i - x_j\|_{\mu+1}^{\mu+1}, \end{aligned} \tag{8.14}$$

i.e., z is always nonpositive, which indicates that V_1 is nonincreasing. Next, we suppose that there exist two agents, x_{i0} and x_{j0}, whose initial connection is to be broken, i.e., $\|x_{i0} - x_{j0}\| \to R$. From (8.11), one obtains that $\int_0^{\|x_{i0} - x_{j0}\|} \alpha \psi(s) s^{\mu} \mathrm{d}s \to +\infty$, so $V_1 \to +\infty$, which contradicts with the fact that V_1 is nonincreasing. Therefore, all initial connections are maintained over time, i.e., the network connectivity is preserved over time. □

Remark 8.5 The distributed optimization with preserved connectivity is not investigated in [5–8, 12]. Moreover, the potentials designed in those references are usually smooth, which do not take unexpected obstacles or noises into account, thereby easing the analysis. In this chapter, we propose a discontinuous potential based connectivity-preserving protocol to maintain all initial connected links while realizing the distributed optimization in a finite time. An interesting direction of future work would be designing a protocol preserving connectivity in a global manner without affecting the distributed optimization process.

Next, we prove that the system (8.1) with protocol (8.10) achieves finite-time consensus. Define $\check{x}_i = x_i - \frac{1}{N} \sum_{j=1}^{N} x_j$. Construct a Lyapunov function candidate $\tilde{V} = \frac{1}{2} \sum_{i=1}^{N} \check{x}_i^T \check{x}_i$. Note that $\Phi(\|\check{x}_i - \check{x}_j\|) = \Phi(\|x_i - x_j\|)$. We denote

$\tilde{\psi}_{ij} = \tilde{\psi}_{ji} \in \Phi(\|\check{x}_i - \check{x}_j\|)$. Let $\mathscr{L}_A$ be the Laplacian matrix of graph $\mathscr{G}([\text{sign}(\tilde{\psi}_{ij}(t = 0))])$, and $\lambda_2(\cdot)$ be the second smallest eigenvalue for a corresponding matrix.

Theorem 8.2 *Under Assumptions 8.1.1–8.2.1, the system (8.1) with protocol (8.10) achieves finite-time consensus with the settling time T_s estimated as*

$$T_s \leq T_1 = \frac{4\tilde{V}^{\frac{1-\mu}{2}}(0)}{\alpha(1-\mu)(4\lambda_2(\mathscr{L}_A))^{\frac{\mu+1}{2}}}. \tag{8.15}$$

Proof According to Definition 8.2, one gets

$$\begin{aligned} \dot{\check{x}}_i \in & -\alpha \sum_{j=1}^{N} \Phi(\|\check{x}_i - \check{x}_j\|)\lceil \check{x}_i - \check{x}_j \rfloor^{\mu} - \gamma \nabla f_i(x_i) + \frac{1}{N}\sum_{j=1}^{N} \gamma \nabla f_j(x_j) \\ = & -\alpha \sum_{j=1}^{N} \Phi(\|\check{x}_i - \check{x}_j\|)\lceil \check{x}_i - \check{x}_j \rfloor^{\mu} - \frac{\gamma\rho}{N}\sum_{j=1}^{N}(\check{x}_i - \check{x}_j). \end{aligned} \tag{8.16}$$

The set-valued Lie derivative of $\tilde{V}$ along (8.16) is

$$\tilde{\mathscr{L}}\tilde{V} = -\alpha \sum_{i,j=1}^{N} \Phi(\|\check{x}_i - \check{x}_j\|)\check{x}_i^T \lceil \check{x}_i - \check{x}_j \rfloor^{\mu} - \frac{\gamma\rho}{N}\sum_{i,j=1}^{N} \check{x}_i^T(\check{x}_i - \check{x}_j). \tag{8.17}$$

For the first term in (8.17), one further obtains

$$\begin{aligned} & -\alpha \sum_{i,j=1}^{N} \Phi(\|\check{x}_i - \check{x}_j\|)\check{x}_i^T \lceil \check{x}_i - \check{x}_j \rfloor^{\mu} \\ = & -\frac{\alpha}{2}\sum_{i,j=1}^{N} \Phi(\|\check{x}_i - \check{x}_j\|)\check{x}_i^T \lceil \check{x}_i - \check{x}_j \rfloor^{\mu} - \frac{\alpha}{2}\sum_{i,j=1}^{N} \Phi(\|\check{x}_j - \check{x}_i\|)\check{x}_j^T \lceil \check{x}_j - \check{x}_i \rfloor^{\mu} \\ = & -\frac{\alpha}{2}\sum_{i,j=1}^{N} \Phi(\|\check{x}_j - \check{x}_i\|)\|\check{x}_i - \check{x}_j\|_{\mu+1}^{\mu+1}. \end{aligned} \tag{8.18}$$

Following the same lines, for the second term in (8.17), one gets

$$-\frac{\gamma\rho}{N}\sum_{i,j=1}^{N} \check{x}_i^T(\check{x}_i - \check{x}_j) = -\frac{\gamma\rho}{2N}\sum_{i,j=1}^{N} \|\check{x}_i - \check{x}_j\|_2^2. \tag{8.19}$$

Combining (8.18)–(8.19), it can be concluded from (8.17) that

$$
\begin{aligned}
\max \widetilde{\mathscr{L}}\tilde{V} &= -\frac{\alpha}{2}\sum_{i,j=1}^{N}\tilde{\psi}_{ij}\left\|\check{x}_i-\check{x}_j\right\|_{\mu+1}^{\mu+1}-\frac{\gamma\rho}{2N}\sum_{i,j=1}^{N}\left\|\check{x}_i-\check{x}_j\right\|_2^2\\
&\le -\frac{\alpha}{2}\sum_{i,j=1}^{N}\tilde{\psi}_{ij}\left\|\check{x}_i-\check{x}_j\right\|_{\mu+1}^{\mu+1}\\
&\le -\frac{\alpha}{2}\sum_{i,j=1}^{N}\left(\tilde{\psi}_{ij}^{\frac{2}{\mu+1}}\left\|\check{x}_i-\check{x}_j\right\|_2^2\right)^{\frac{\mu+1}{2}}\\
&\le -\frac{\alpha}{2}\left(\sum_{i,j=1}^{N}\tilde{\psi}_{ij}^{\frac{2}{\mu+1}}\left\|\check{x}_i-\check{x}_j\right\|_2^2\right)^{\frac{\mu+1}{2}},
\end{aligned}
$$

where Lemma 3.3 is used to obtain the last inequality. Let $\mathscr{L}_B(t)$ and $\mathscr{L}_C(t)$ denote the Laplacian matrix of graph $\mathscr{G}([\tilde{\psi}_{ij}(t)])$ and $\mathscr{G}([\tilde{\psi}_{ij}^{2/(\mu+1)}(t)])$, respectively. From Lemma 3.1 in [13], one obtains that $\lambda_2(\mathscr{L}_B(t)) \ge \lambda_2(\mathscr{L}_A)$. Similarly, it is ready to obtain $\lambda_2(\mathscr{L}_C(t)) \ge \lambda_2(\mathscr{L}_A)$ due to the second property in Sect. 8.2.2. Note that $\sum_{i=1}^{N}\sum_{j=1}^{N}\tilde{\psi}_{ij}^{2/(\mu+1)}\left\|\check{x}_i-\check{x}_j\right\|_2^2 = 2\check{x}^T(\mathscr{L}_C(t)\otimes I_m)\check{x} \ge 2\lambda_2(\mathscr{L}_C(t))\check{x}^T\check{x}$, where $\check{x}=[\check{x}_1,\check{x}_2,\ldots,\check{x}_N]^T$ [14]. Therefore,

$$
\begin{aligned}
\max \widetilde{\mathscr{L}}\tilde{V} &\le -\frac{\alpha}{2}\left(2\lambda_2(\mathscr{L}_C(t))\check{x}^T\check{x}\right)^{\frac{\mu+1}{2}}\\
&\le -\frac{\alpha}{2}\left(4\lambda_2(\mathscr{L}_A)\tilde{V}\right)^{\frac{\mu+1}{2}}\\
&= -\frac{\alpha}{2}\left(4\lambda_2(\mathscr{L}_A)\right)^{\frac{\mu+1}{2}}\tilde{V}^{\frac{\mu+1}{2}}.
\end{aligned}
\tag{8.20}
$$

Although $\dot{\tilde{V}}$ is discontinuous at some time instants, it is Riemann integrable as $\tilde{V}$ is bounded and the set of points of discontinuity has measure zero [15]. Therefore, from (8.20), one gets

$$
\tilde{V}(t+h)-\tilde{V}(t)=\int_t^{t+h}\dot{\tilde{V}}\mathrm{d}\tau \le -\frac{\alpha h}{2}(4\lambda_2(\mathscr{L}_A))^{\frac{\mu+1}{2}}\min_{\tau\in[t,t+h]}\tilde{V}^{\frac{\mu+1}{2}}(\tau).
$$

where $h \to 0^+$. From the definition of Dini derivative, one further gets

$$
D^*\tilde{V}(t) \le -\frac{\alpha}{2}(4\lambda_2(\mathscr{L}_A))^{\frac{\mu+1}{2}}\tilde{V}^{\frac{\mu+1}{2}}(t),
$$

Invoking Theorem 2.2, it can be concluded that $\tilde{V}=0$ within a settling time $T_1 = 4\tilde{V}^{(1-\mu)/2}(0)/(\alpha(1-\mu)(4\lambda_2(\mathscr{L}_A))^{(\mu+1)/2})$. Therefore, the state agreement is realized in a finite time T_1, which completes the proof. □

Theorem 8.3 *Under Assumptions 8.1.1–8.2.1, the system (8.1) with protocol (8.10) achieves finite-time consensus within T_1, and minimizes the overall cost function in (8.2) when $t \to \infty$.*

Proof From Theorem 8.2, the states of all agents reach a consensus for $t \geq T_1$, i.e., $x_i(t) = \frac{1}{N}\sum_{j=1}^{N} x_j(t) := x_c(t), \forall i \in \Gamma$. Then

$$\begin{aligned}\dot{x}_c(t) &= -\frac{\alpha}{N}\sum_{i,j=1}^{N} \psi(\|x_i - x_j\|)\lceil x_i - x_j \rfloor^{\mu} - \frac{\gamma}{N}\sum_{i=1}^{N} \nabla f_i(x_i)\\ &= -\frac{\gamma}{N}\sum_{i=1}^{N} \nabla f_i(x_c),\end{aligned} \tag{8.21}$$

where the symmetric property of undirected graphs is used. Moreover, one obtains

$$\begin{aligned}\sum_{i=1}^{N} \dot{f}_i(x_c) &= \sum_{i=1}^{N} \nabla f_i^T(x_c)\dot{x}_c\\ &= -\frac{\gamma}{N}\sum_{i=1}^{N} \nabla f_i^T(x_c)\sum_{j=1}^{N} \nabla f_j(x_c)\\ &= -\frac{\gamma}{N}\left(\sum_{i=1}^{N} \nabla f_i(x_c)\right)^T \sum_{j=1}^{N} \nabla f_j(x_c).\end{aligned} \tag{8.22}$$

Under Assumption 8.1.1, one obtains that $\sum_{i=1}^{N} f_i(x)$ is lower bounded. As a result, it can be concluded that $\lim_{t\to\infty}\sum_{i=1}^{N} \nabla f_i(x_c) = 0$. By using Lemma 7.1, $x = x_c$ minimizes the differentiable convex function $\sum_{i=1}^{N} f_i(x)$ when $t \to \infty$. □

8.3 Distributed Optimization with Network Connectivity and Fixed-Time Consensus

The estimated settling time in (8.15) relies on initial states. However, in practical scenarios, the knowledge of initial conditions is sometimes unavailable or inaccurate, which limits the implementation of protocol (8.10). To address this problem, based on protocol (8.10), we develop a distributed protocol for the ith agent, i.e.,

$$u_i = -\alpha \sum_{j=1}^{N} \psi(\|x_i - x_j\|)\lceil x_i - x_j \rfloor^{\mu}$$

$$-\beta \sum_{j=1}^{N} \psi(\|x_i - x_j\|)\lceil x_i - x_j \rfloor^{\nu} - \gamma \nabla f_i(x_i), \tag{8.23}$$

where $\beta > 0$, and $\nu > 1$.

Remark 8.6 Compared to protocol (8.10), protocol (8.23) includes an additional fractional power term with an index $\nu > 1$. On one hand, this power term ensures that the state agreement, namely $\|x_i(t) - x_j(t)\| = 0$ for all $i, j \in \Gamma$, is realized in a fixed time, which provides convergence information without dependence on initial states. On the other hand, if the initial state disagreement is large, the initial control effort for u_i in (8.23) becomes relatively large compared to that in protocol (8.10). If letting $\beta = 0$, protocol (8.23) can be reduced to the form of (8.10).

Theorem 8.4 *Under Assumption 8.1.1, for the system (8.1) with protocol (8.23), the network connectivity is preserved over time.*

Proof Construct the following nonnegative function

$$V_2 = \frac{1}{2} \sum_{i,j=1}^{N} \int_0^{\|x_i - x_j\|} \psi(s)(\alpha s^{\mu} + \beta s^{\nu}) \mathrm{d}s, \tag{8.24}$$

and the remaining proof follows a similar procedure of Theorem 8.1, thus omitted. □

Theorem 8.5 *Under Assumptions 8.1.1–8.2.1, the system (8.1) with protocol (8.23) achieves fixed-time consensus with the settling time T_s estimated as*

$$T_s \leq T_2 = \frac{4}{\alpha(1-\mu)(4\lambda_2(\mathscr{L}_A))^{\frac{\mu+1}{2}}} + \frac{4}{\beta m^{\frac{1-\nu}{2}} N^{1-\nu}(\nu-1)(4\lambda_2(\mathscr{L}_A))^{\frac{\nu+1}{2}}}. \tag{8.25}$$

Proof According to Definition 8.2, one gets

$$\dot{\check{x}}_i \in -\alpha \sum_{j=1}^{N} \Phi(\|\check{x}_i - \check{x}_j\|)\lceil \check{x}_i - \check{x}_j \rfloor^{\mu}$$
$$-\beta \sum_{j=1}^{N} \Phi(\|\check{x}_i - \check{x}_j\|)\lceil \check{x}_i - \check{x}_j \rfloor^{\nu} \tag{8.26}$$
$$-\frac{\gamma\rho}{N} \sum_{j=1}^{N} (\check{x}_i - \check{x}_j). \tag{8.27}$$

The set-valued Lie derivative of $\tilde{V} = \frac{1}{2} \sum_{i=1}^{N} \check{x}_i^T \check{x}_i$ along (8.26) is

$$\begin{aligned}\widetilde{\mathscr{L}}\tilde{V} =& -\alpha \sum_{i,j=1}^{N} \Phi(\|\check{x}_i - \check{x}_j\|)\check{x}_i^T \lceil \check{x}_i - \check{x}_j \rfloor^{\mu} \\ & - \beta \sum_{i,j=1}^{N} \Phi(\|\check{x}_i - \check{x}_j\|)\check{x}_i^T \lceil \check{x}_i - \check{x}_j \rfloor^{\nu} \quad (8.28) \\ & - \frac{\gamma\rho}{N} \sum_{i,j=1}^{N} \check{x}_i^T (\check{x}_i - \check{x}_j). \quad (8.29)\end{aligned}$$

Similar to (8.18) and (8.19), it can be concluded from (8.28) that

$$\begin{aligned}\max \widetilde{\mathscr{L}}\tilde{V} =& -\frac{\alpha}{2} \sum_{i,j=1}^{N} \tilde{\psi}_{ij} \|\check{x}_i - \check{x}_j\|_{\mu+1}^{\mu+1} - \frac{\beta}{2} \sum_{i,j=1}^{N} \tilde{\psi}_{ij} \|\check{x}_i - \check{x}_j\|_{\nu+1}^{\nu+1} - \frac{\gamma\rho}{2N} \sum_{i,j=1}^{N} \|\check{x}_i - \check{x}_j\|_2^2 \\ \leq& -\frac{\alpha}{2} \sum_{i,j=1}^{N} \left(\tilde{\psi}_{ij}^{\frac{2}{\mu+1}} \|\check{x}_i - \check{x}_j\|_2^2 \right)^{\frac{\mu+1}{2}} - \frac{\beta m^{\frac{1-\nu}{2}}}{2} \sum_{i,j=1}^{N} \left(\tilde{\psi}_{ij}^{\frac{2}{\nu+1}} \|\check{x}_i - \check{x}_j\|_2^2 \right)^{\frac{\nu+1}{2}} \\ \leq& -\frac{\alpha}{2} \left(\sum_{i,j=1}^{N} \tilde{\psi}_{ij}^{\frac{2}{\mu+1}} \|\check{x}_i - \check{x}_j\|_2^2 \right)^{\frac{\mu+1}{2}} \\ & - \frac{\beta m^{\frac{1-\nu}{2}} N^{1-\nu}}{2} \left(\sum_{i,j=1}^{N} \tilde{\psi}_{ij}^{\frac{2}{\nu+1}} \|\check{x}_i - \check{x}_j\|_2^2 \right)^{\frac{\nu+1}{2}}. \quad (8.30)\end{aligned}$$

Let $\mathscr{L}_D(t)$ be the Laplacian matrix of graph $\mathscr{G}([\tilde{\psi}_{ij}^{2/(\nu+1)}(t)])$. One can obtain that $\lambda_2(\mathscr{L}_D(t)) \geq \lambda_2(\mathscr{L}_A)$. Moreover, $\sum_{i=1}^{N} \sum_{j=1}^{N} \tilde{\psi}_{ij}^{2/(\nu+1)} \|\check{x}_i - \check{x}_j\|_2^2 = 2\check{x}^T (\mathscr{L}_D(t) \otimes I_m)\check{x} \geq 2\lambda_2(\mathscr{L}_D(t))\check{x}^T \check{x}$. Therefore,

$$\begin{aligned}\max \widetilde{\mathscr{L}}\tilde{V} \leq& -\frac{\alpha}{2}(4\lambda_2(\mathscr{L}_A))^{\frac{\mu+1}{2}} \tilde{V}^{\frac{\mu+1}{2}} \\ & - \frac{\beta m^{\frac{1-\nu}{2}} N^{1-\nu}}{2}(4\lambda_2(\mathscr{L}_A))^{\frac{\nu+1}{2}} \tilde{V}^{\frac{\nu+1}{2}}. \quad (8.31)\end{aligned}$$

From (8.31), one gets

$$\begin{aligned}\tilde{V}(t+h) - \tilde{V}(t) =& \int_t^{t+h} \dot{\tilde{V}} \mathrm{d}\tau \\ \leq& -\frac{\alpha h}{2}(4\lambda_2(\mathscr{L}_A))^{\frac{\mu+1}{2}} \min_{\tau \in [t,t+h]} \tilde{V}^{\frac{\mu+1}{2}}(\tau) \\ & - \frac{\beta h m^{\frac{1-\nu}{2}} N^{1-\nu}}{2}(4\lambda_2(\mathscr{L}_A))^{\frac{\nu+1}{2}} \min_{\tau \in [t,t+h]} \tilde{V}^{\frac{\nu+1}{2}}(\tau). \quad (8.32)\end{aligned}$$

From the definition of Dini derivative, one further obtains

$$D^+\tilde{V}(t) \leq -\frac{\alpha}{2}(4\lambda_2(\mathscr{L}_A))^{\frac{\mu+1}{2}}\tilde{V}^{\frac{\mu+1}{2}}(t)$$
$$-\frac{\beta m^{\frac{1-\nu}{2}}N^{1-\nu}}{2}(4\lambda_2(\mathscr{L}_A))^{\frac{\nu+1}{2}}\tilde{V}^{\frac{\nu+1}{2}}(t). \tag{8.33}$$

Using Theorem 2.4 and *Comparison Principle*, it can be concluded that $\tilde{V} = 0$ within a settling time T_2. Therefore, the consensus is achieved in a fixed time T_2, which completes the proof. □

Theorem 8.6 *Under Assumptions 8.1.1–8.2.1, the system (8.1) with protocol (8.23) achieves fixed-time consensus within* T_2, *and minimizes the overall cost function in (8.2) as* $t \to \infty$.

Proof Since the proof follows a similar procedure of Theorem 8.3, we omit the procedures. □

Remark 8.7 With protocol (8.10) or (8.23), the estimated settling time depends on global information of the interaction graph, i.e., $\lambda_2(\mathscr{L}_A)$. How to obtain $\lambda_2(\mathscr{L})$ in a fully distributed manner without affecting the distributed optimization process deserves further investigation.

Remark 8.8 The protocols (8.10) and (8.23) are proposed based on Assumption 8.1.1 requiring the cost function $f_i(x)$ to be quadratic. It should be pointed out that the results obtained in Sects. 8.2 and 8.3 can be applied to other convex cost functions. In what follows, we first provide the steps to show that Theorems 8.1–8.3 and Corollaries 8.1–8.6 can hold with a fourth-order cost function as an example. Then we discuss the forms of cost functions that can be optimized within the framework proposed in the current chapter. To begin with, suppose that the cost function is

$$f_i(x) = ax^4 + bx^2 + cx + d_i, \tag{8.34}$$

where $a > 0, b > 0, c \in \mathbb{R}$, and $d_i \in \mathbb{R}^m$. Obviously, (8.34) has a more general form than the quadratic form. To simplify the analysis, we consider the one-dimensional case, i.e., $m = 1$. Then (8.14) becomes

$$\begin{aligned} z = &-\sum_{i=1}^{N}\Big|\sum_{j=1}^{N}\alpha\psi_{ij}\lceil x_i - x_j\rfloor^{\mu}\Big|^2 + \sum_{i,j=1}^{N}\alpha\psi_{ij}\lceil x_i - x_j\rfloor^{\mu}(-4\gamma a x_i^3 - 2\gamma b x_i - \gamma c) \\ = &-\sum_{i=1}^{N}\Big|\sum_{j=1}^{N}\alpha\psi_{ij}\lceil x_i - x_j\rfloor^{\mu}\Big|^2 - \gamma\sum_{i,j=1}^{N}\alpha\psi_{ij}\lceil x_i - x_j\rfloor^{\mu}(2ax_i^3 - 2ax_j^3 + bx_i - bx_j) \\ = &-\sum_{i=1}^{N}\Big|\sum_{j=1}^{N}\alpha\psi_{ij}\lceil x_i - x_j\rfloor^{\mu}\Big|^2 - 2\gamma a\sum_{i,j=1}^{N}\alpha\psi_{ij}|x_i - x_j|^{\mu+1}(x_i^2 + x_i x_j + x_j^2) \\ &- \gamma b\sum_{i,j=1}^{N}\alpha\psi_{ij}|x_i - x_j|^{\mu+1}. \end{aligned} \tag{8.35}$$

Therefore, z is always nonpositive. As a result, Theorem 8.1 holds with $m = 1$. Following a similar procedure, it can be obtained that Theorems 8.2–8.3 and Corollaries 8.1–8.6 hold as well with $m = 1$. From the above analysis, one can conclude that the cost function $f_i(x)$ is not restricted to the quadratic form. In fact, other forms of cost functions including the illustrated fourth-order one can also be optimized while network connectivity is preserved over time in the presence of obstacles. However, it should be pointed out that it is tricky to select compatible forms to maintain the network connectivity, achieve the finite-time consensus, and solve the distributed optimization problem all together. One has to obtain the nonpositive property of z as in (8.14) and (8.35). In addition, one needs to make sure that the term of $\nabla f_i(x_i)$ is nonpositive during the calculation of $\widetilde{\mathscr{L}}\tilde{V}$ as in (8.19).

8.4 Numerical Examples

We provide examples to show the effectiveness of the proposed protocols. It is assumed that $N = 5$ and $m = 2$. The interaction rule described in Sect. 8.2.2 is implemented in this section.

We first consider that the sensing range is $R = 2.5$. An artificial potential $\psi(s)$ satisfying the generalized properties described in Sect. 8.2.2 is selected as

$$\psi(s) = \begin{cases} \frac{R^2}{R^2 - s^2}, & 0 \le s < \frac{R}{2}; \\ \frac{0.75R^2}{R^2 - s^2}, & \frac{R}{2} \le s < R; \\ 0, & s \ge R. \end{cases} \tag{8.36}$$

A graph shown in Fig. 1.5 is used as initial configuration, where the states for agents are $\{(-4, -1), (-1.732, -1), (0, 0), (0, 2), (1.732, -1)\}$. Initially, $\|x_1 - x_2\| = 2.27$, $\|x_2 - x_3\| = 2$, $\|x_3 - x_4\| = 2$, $\|x_3 - x_5\| = 2$, corresponding links are built between agents with relative distance smaller than R. Moreover, one obtains that $\lambda_2(\mathscr{L}_A) = 0.5188$. The local cost function for the ith agent is set as $f_i(x_i) = (x_{i1} + 0.5)^2 + (x_{i2} - 0.5)^2 + i$. It can be easily calculated that the minima of the overall cost function $\sum_{i=1}^{5} f_i(x_i)$ is 15. Select $\alpha = 5$, $\gamma = 2$, and $\mu = 0.2$. Based on Theorem 8.2, one obtains $T_1 = 1.79$ s. Implementing protocol (8.10) with $N = 5$, Figs. 8.1 and 8.2 show that the settling time is about $t = 0.5$ s, which is smaller than T_1. The evolution of $\sum_{i=1}^{5} f_i(x_i)$ is shown in Fig. 8.3, which approaches to the optimal value 15. In Fig. 8.4, the relative distance of initial connected agents are smaller than R over time, which demonstrates the effectiveness of the connectivity preservation property of protocol (8.10).

In order to highlight the connectivity preservation capability of the proposed protocols, we consider that agent 3 is suffering from an unexpected obstacle with a negative repelling potential when $0.1 \le t \le 0.2$ s, i.e., $i \in \{1, 2, 4, 5\}$,

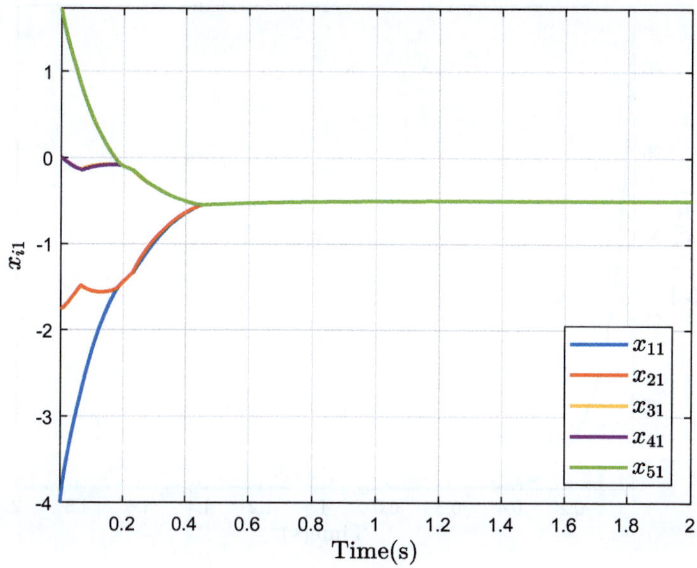

Fig. 8.1 State x_{i1}

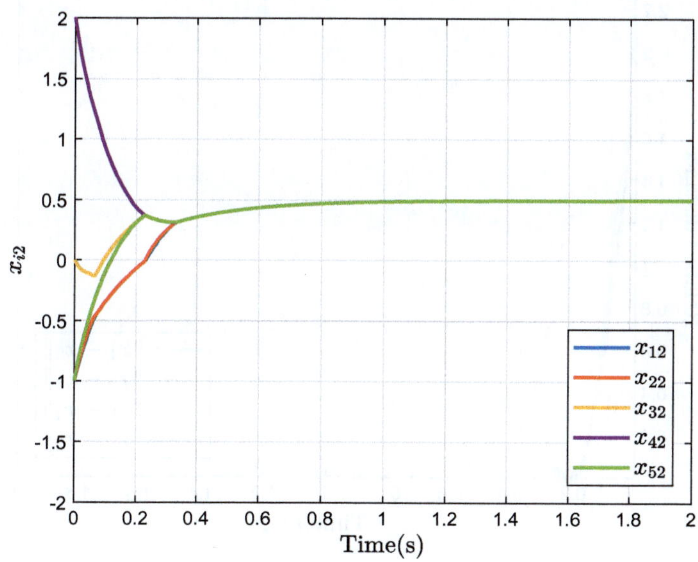

Fig. 8.2 State x_{i2}

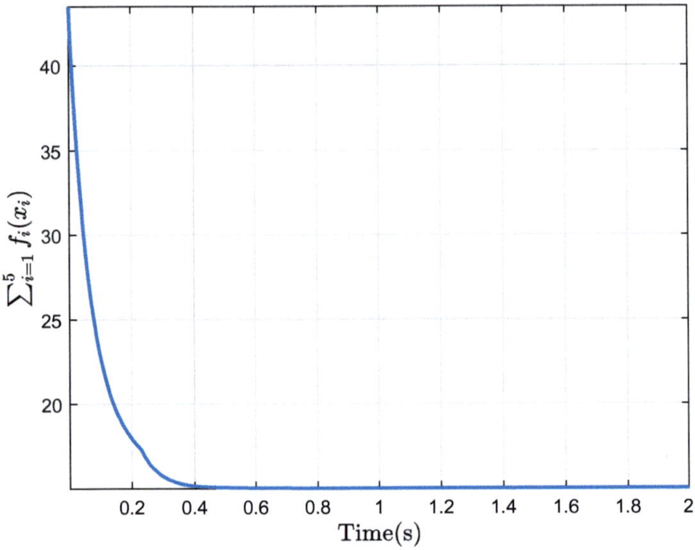

Fig. 8.3 The overall cost function

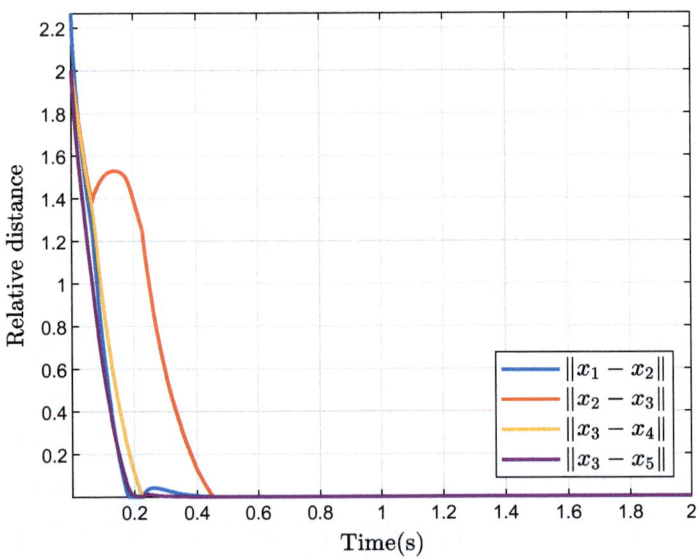

Fig. 8.4 Connectivity preservation with protocol (8.10)

$$u_i(t) = -\alpha \sum_{j=1}^{5} \psi(\|x_i - x_j\|)\lceil x_i - x_j \rfloor^{0.2} - \gamma \nabla f_i(x_i), \ t > 0, \tag{8.37}$$

and for the agent 3,

$$u_3(t) = \begin{cases} -\alpha \sum_{j=1}^{5} \psi(\|x_i - x_j\|)\lceil x_i - x_j \rfloor^{0.2} - \gamma \nabla f_i(x_i) - \begin{bmatrix} \sin(2\pi t) - 60 \\ \cos(3\pi t) - 60 \end{bmatrix} \\ \qquad \text{when } 0.1 \le t \le 0.2; \\ -\alpha \sum_{j=1}^{5} \psi(\|x_i - x_j\|)\lceil x_i - x_j \rfloor^{0.2} - \gamma \nabla f_i(x_i) \\ \qquad \text{when } 0 < t < 0.1 \cup t > 0.2, \end{cases} \tag{8.38}$$

where $\psi(\cdot)$ is defined in (8.36). The simulation results in Figs. 8.5, 8.6, 8.7, and 8.8 show that the consensus is achieved at $t = 0.4$ s, and the initial connected links ($\|x_1 - x_2\|$, $\|x_2 - x_3\|$, $\|x_3 - x_4\|$, and $\|x_3 - x_5\|$) have lengths smaller than 2.5 over time.

To make a comparison, a conventional finite-time protocol is selected as, $i \in \{1, 2, 4, 5\}$,

$$u_i(t) = -\alpha \sum_{j=1}^{5} a_{ij}\lceil x_i - x_j \rfloor^{0.2} - \gamma \nabla f_i(x_i), \ t > 0, \tag{8.39}$$

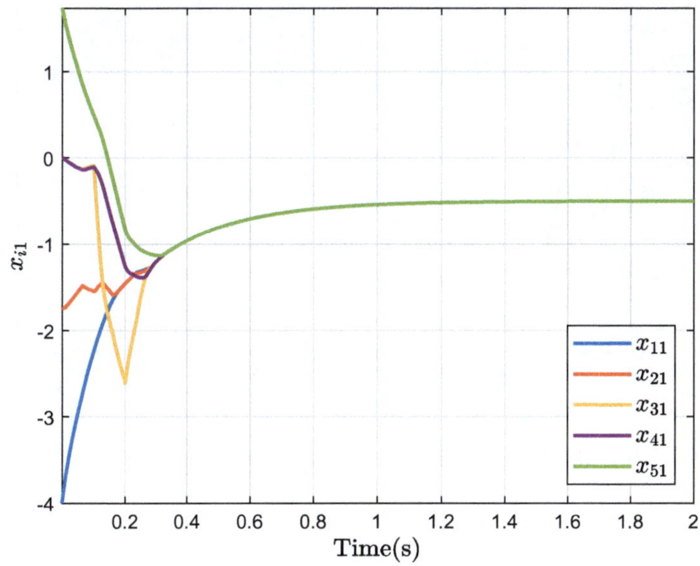

Fig. 8.5 State x_{i1}

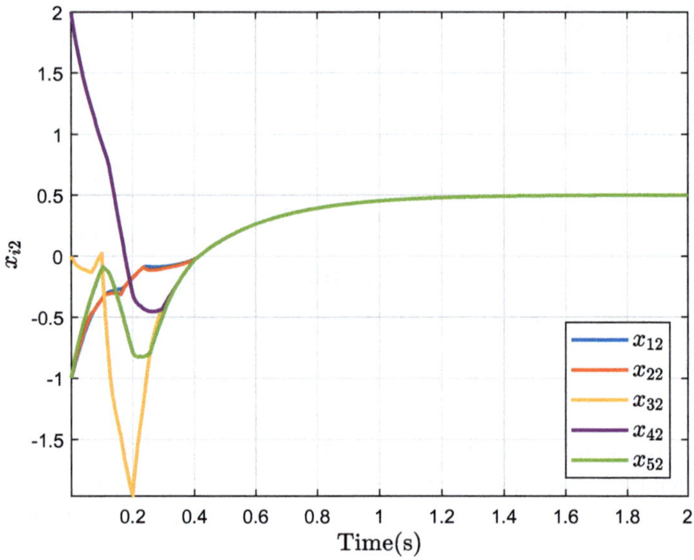

Fig. 8.6 State x_{i2}

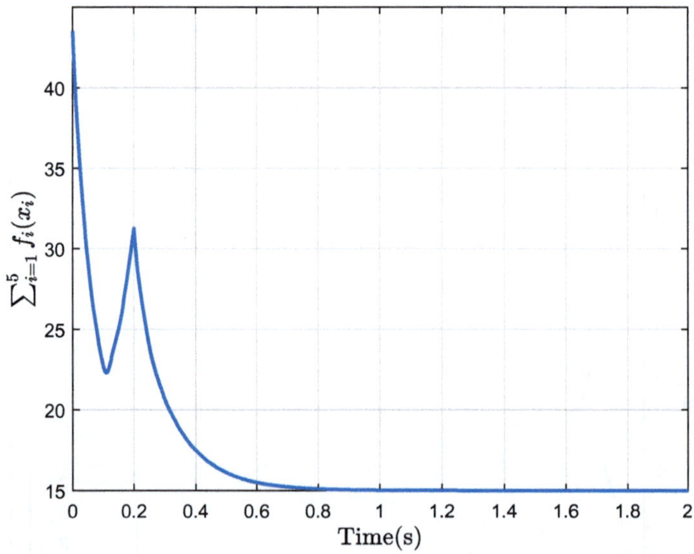

Fig. 8.7 The overall cost function

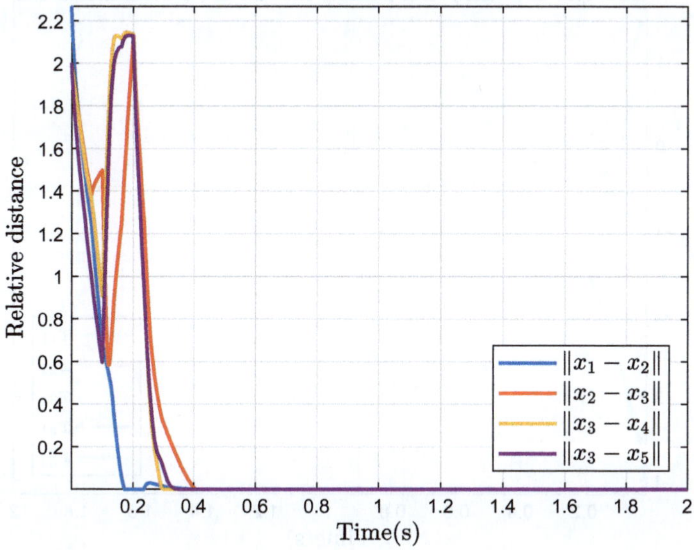

Fig. 8.8 Connectivity preservation with protocol (8.37)–(8.38)

and for the agent 3,

$$u_3(t) = \begin{cases} -\alpha \sum_{j=1}^{5} a_{ij} \lceil x_i - x_j \rfloor^{0.2} - \gamma \nabla f_i(x_i) - \begin{bmatrix} \sin(2\pi t) - 60 \\ \cos(3\pi t) - 60 \end{bmatrix} \\ \qquad\qquad \text{when } 0.1 \leq t \leq 0.2; \\ -\alpha \sum_{j=1}^{5} a_{ij} \lceil x_i - x_j \rfloor^{0.2} - \gamma \nabla f_i(x_i) \\ \qquad\qquad \text{when } 0 < t < 0.1 \cup t > 0.2, \end{cases} \tag{8.40}$$

where a_{ij} is corresponding to a fixed $\psi(s) = 1$, i.e., the connectivity strength is unweighted and state independent. The simulation results in Figs. 8.9, 8.10, 8.11, and 8.12 show that the convergence speed becomes slower and exact consensus is not achieved (only asymptotic consensus instead), and the initial connected links ($\|x_2 - x_3\|$, $\|x_3 - x_4\|$, and $\|x_3 - x_5\|$) are broken at $t = 0.13$ s. Although the overall cost function is optimized, the interaction among agents is lost (agents become isolated) after $t = 0.13$ s. As a result, these agents cannot cooperatively accomplish globally allocated tasks.

To show the effectiveness of the fixed-time protocol (8.23), we then change R to 250, and change the initial states of agents to

$$\{(-400, -100), (-173.2, -100), (0, 0), (0, 200), (173.2, -100)\}.$$

Select $\beta = 3$ and $\nu = 1.3$. Based on Theorem 8.2, one obtains $T_2 = 20.05$ s. Implementing protocol (8.23) with $N = 5$, Figs. 8.13 and 8.14 show that the settling time

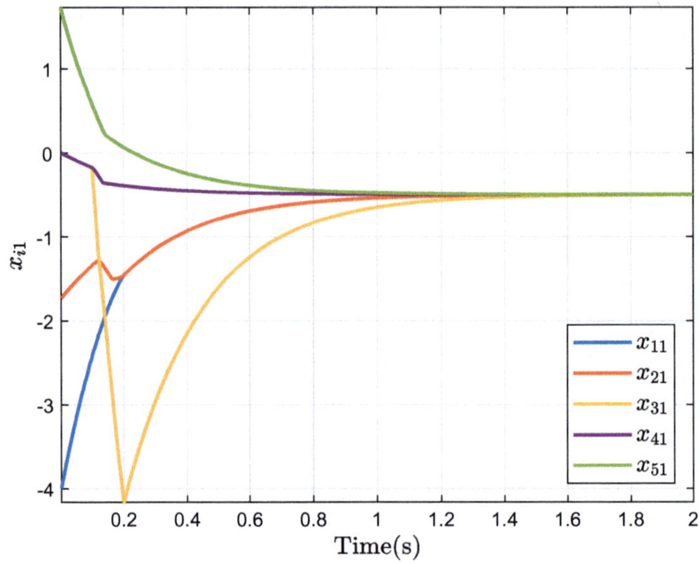

Fig. 8.9 State x_{i1}

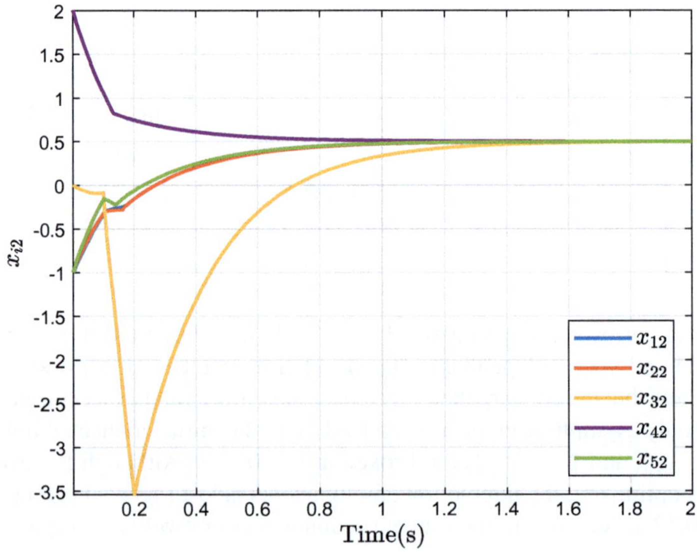

Fig. 8.10 State x_{i2}

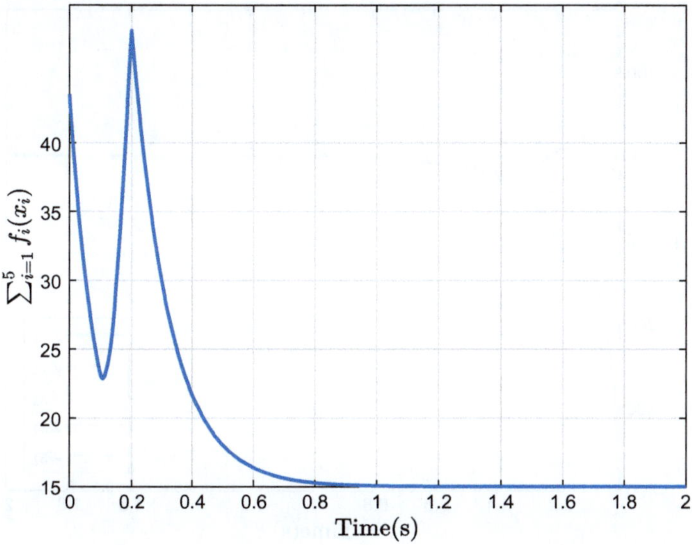

Fig. 8.11 The overall cost function

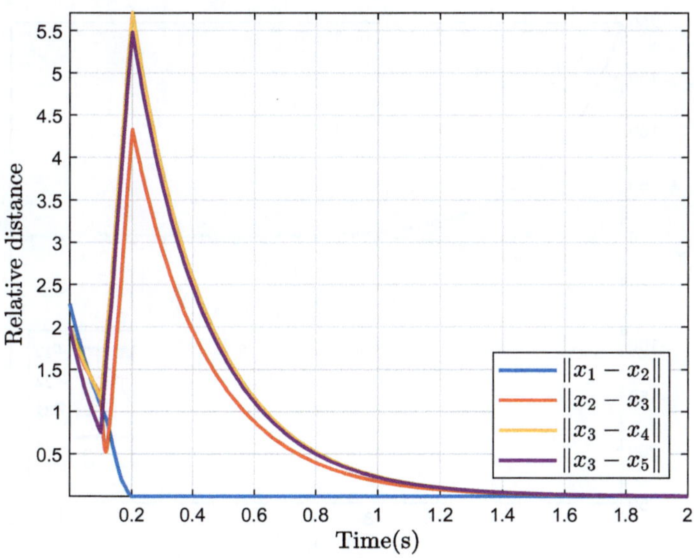

Fig. 8.12 Links broken with protocol (8.39)–(8.40)

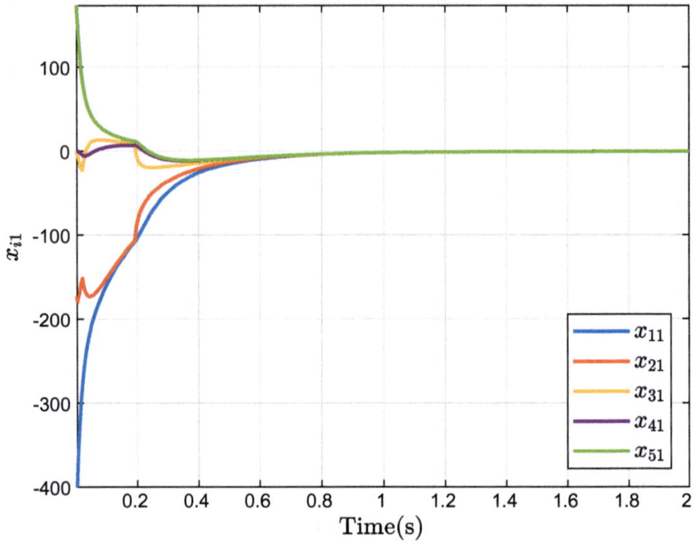

Fig. 8.13 State x_{i1}

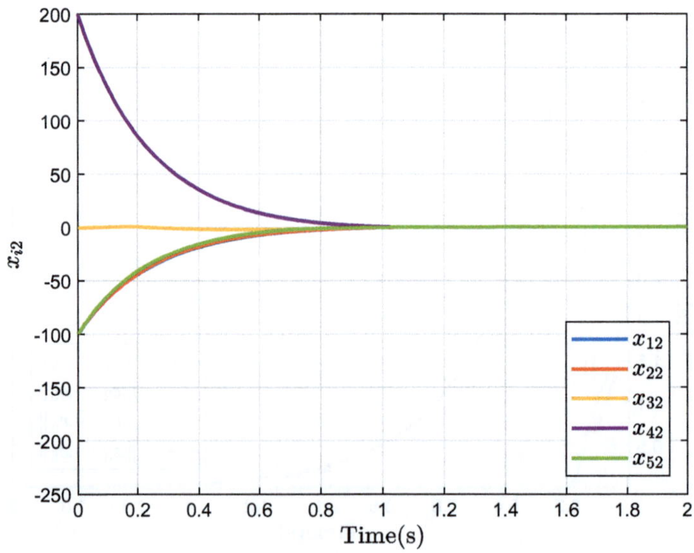

Fig. 8.14 State x_{i2}

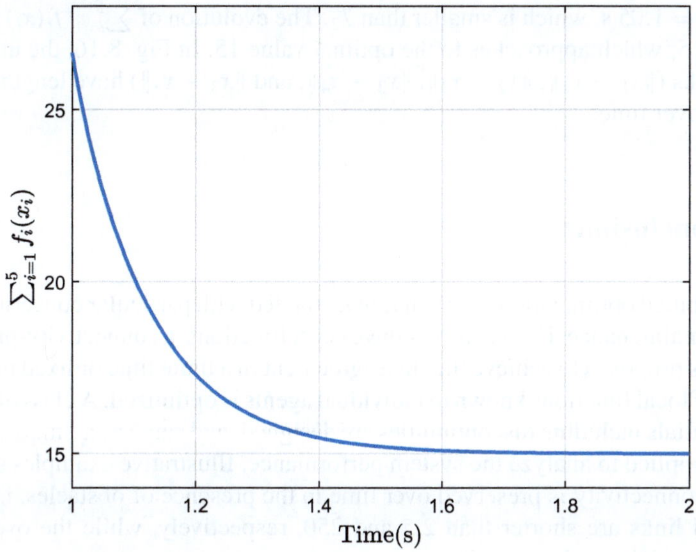

Fig. 8.15 The overall cost function

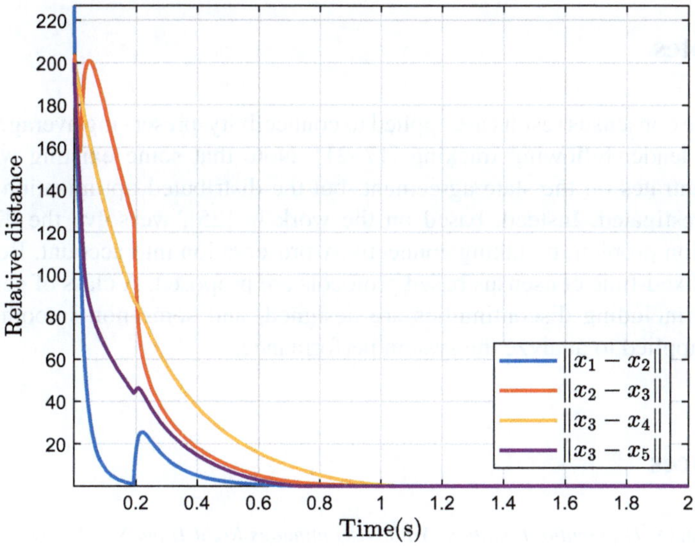

Fig. 8.16 Connectivity preservation with protocol (8.23)

is about $t = 1.05$ s, which is smaller than T_2. The evolution of $\sum_{i=1}^{5} f_i(x_i)$ is shown in Fig. 8.15, which approaches to the optimal value 15. In Fig. 8.16, the initial connected links ($\|x_1 - x_2\|$, $\|x_2 - x_3\|$, $\|x_3 - x_4\|$, and $\|x_3 - x_5\|$) have lengths smaller than 250 over time.

8.5 Conclusions

The distributed optimization problem is investigated with particular concerns on connectivity maintenance. For a class of convex cost functions, a connectivity-preserving protocol is proposed to achieve the state agreement in a finite time or fixed time while the sum of local functions known to individual agents is optimized. A class of generalized potentials including discontinuities are designed, and some non-smooth analysis tools are applied to analyze the system performance. Illustrative examples show that network connectivity is preserved over time in the presence of obstacles, i.e., initial connected links are shorter than 2.5 and 250, respectively, while the overall cost function is minimized.

8.6 Notes

Fixed-time consensus results are applied to connectivity-preserving average tracking [16], and leader-following tracking [17–21]. Note that some existing work [16–24] concentrates on the state agreement, but the distributed optimization problem is not investigated. Instead, based on the work in [25], we solve the distributed optimization problem by taking connectivity preservation into account. Both finite-time and fixed-time consensus based protocols are proposed. A class of generalized potentials including discontinuities are designed, and some non-smooth analysis tools are applied to analyze the system performance.

References

1. A. Filippov, *Differential Equations With Discontinuous Right-Hand Side* (Kluwer Academic, Dordrecht, 1988)
2. D. Shevitz, B. Paden, Lyapunov stability theory of nonsmooth systems. IEEE Trans. Autom. Control **39**(9), 1910–1914 (1994)
3. A. Bacciotti, F. Ceragioli, Stability and stabilization of discontinuous systems and nonsmooth Lyapunov functions. ESAIM: Control Optim. Calculus Var. **4**, 361–376 (1999)
4. A. Goldsmith, *Wireless Communications* (Cambridge University Press, Cambridge, 2005)
5. M. Ji, M. Egerstedt, Distributed coordination control of multiagent systems while preserving connectedness. IEEE Trans. Robot. **23**(4), 693–703 (2007)
6. D.V. Dimarogonas, K.J. Kyriakopoulos, Connectedness preserving distributed swarm aggregation for multiple kinematic robots. IEEE Trans. Robot. **24**(5), 1213–1223 (2008)

7. M.M. Zavlanos, G.J. Pappas, Distributed connectivity control of mobile networks. IEEE Trans. Robot. **24**(6), 1416–1428 (2008)
8. R.K. Williams, G.S. Sukhatme, Constrained interaction and coordination in proximity-limited multi-agent systems. IEEE Trans. Robot. **29**(4), 930–944 (2013)
9. L. Wang, F. Xiao, Finite-time consensus problems for networks of dynamic agents. IEEE Trans. Autom. Control **55**(4), 950–955 (2010)
10. G. Chen, F.L. Lewis, L. Xie, Finite-time distributed consensus via binary control protocols. Automatica **47**(9), 1962–1968 (2011)
11. F. Xiao, L. Wang, J. Chen, Y. Gao, Finite-time formation control for multi-agent systems. Automatica **45**(11), 2605–2611 (2009)
12. G. Notarstefano, K. Savla, F. Bullo, A. Jadbabaie, Maintaining limited-range connectivity among second-order agents, In *Proceedings of the American Control Conference (ACC)* (Minnesota, USA, Minneapolis, 2006), pp. 2124–2129
13. Y. Cao, W. Ren, D.W. Casbeer, C. Schumacher, Finite-time connectivity-preserving consensus of networked nonlinear agents with unknown Lipschitz terms. IEEE Trans. Autom. Control **61**(6), 1700–1705 (2016)
14. R. Olfati-Saber, R.M. Murray, Consensus problems in networks of agents with switching topology and time-delays. IEEE Trans. Autom. Control **49**(9), 1520–1533 (2004)
15. T.M. Apostol, *Mathematical Analysis* (Addison-Wesley, Boston, 1974)
16. H. Hong, W. Yu, X. Yu, G. Wen, A. Alsaedi, Fixed-time connectivity-preserving distributed average tracking for multi-agent systems. IEEE Trans. Circuits Syst. II: Express Briefs **64**(10), 1192–1196 (2017)
17. B. Ning, Q.-L. Han, Prescribed finite-time consensus tracking for multi-agent systems with nonholonomic chained-form dynamics. IEEE Trans. Autom. Control **64**(4), 1686–1693 (2019)
18. M. Defoort, A. Polyakov, G. Demesure, M. Djemai, K. Veluvolu, Leader-follower fixed-time consensus for multi-agent systems with unknown non-linear inherent dynamics. IET Control Theory Appl. **9**(14), 2165–2170 (2015)
19. Z. Zuo, B. Tian, M. Defoort, Z. Ding, Fixed-time consensus tracking for multi-agent systems with high-order integrator dynamics. IEEE Trans. Autom. Control **63**(2), 563–570 (2018)
20. J. Fu, J. Wang, Finite-time consensus for multi-agent systems with globally bounded convergence time under directed communication graphs. Int. J. Control **90**(9), 1807–1817 (2017)
21. J. Ni, L. Liu, C. Liu, J. Liu, Fixed-time leader-following consensus for second-order multiagent systems with input delay. IEEE Trans. Ind. Electron. **64**(11), 8635–8646 (2017)
22. A. Polyakov, Nonlinear feedback design for fixed-time stabilization of linear control systems. IEEE Trans. Autom. Control **57**(8), 2106–2110 (2012)
23. Z. Zuo, Nonsingular fixed-time consensus tracking for second-order multi-agent networks. Automatica **54**, 305–309 (2015)
24. Z. Zuo, Q.-L. Han, B. Ning, X. Ge, X.-M. Zhang, An overview of recent advances in fixed-time cooperative control of multi-agent systems. IEEE Trans. Ind. Inf. **14**(6), 2322–2334 (2018)
25. B. Ning, Q.-L. Han, Z. Zuo, Distributed optimization of multiagent systems with preserved network connectivity. IEEE Trans. Cybern., to be published, https://doi.org/10.1109/TCYB.2018.2856508

Index

Z. Zuo et al., *Fixed-Time Cooperative Control of Multi-Agent Systems*,
https://doi.org/10.1007/978-3-030-20279-8

MIX
Papier aus verantwortungsvollen Quellen
Paper from responsible sources
FSC® C105338

If you have any concerns about our products,
you can contact us on
ProductSafety@springernature.com

In case Publisher is established outside the EU,
the EU authorized representative is:
Springer Nature Customer Service Center GmbH
Europaplatz 3, 69115 Heidelberg, Germany

Printed by Libri Plureos GmbH
in Hamburg, Germany